Engineering Technologies

Engineering Technologies covers the three mandatory units of the EAL Level 2 Diploma in Engineering and Technology.

- Each compulsory unit is covered in detail with activities, practice exercises and examples where relevant.
- Review questions are provided at the end of each chapter and a sample multiple-choice examination paper is included at the end of the book.
- Contains expert advice that has been written in collaboration with EAL to ensure that it covers what learners need to know.
- Answers to selected questions in the book, together with other supporting resources, can be found on the book's accompanying website. Numerical answers are provided in the book itself.

Written specifically for the EAL Level 2 Diploma in Engineering and Technology, this book covers the three mandatory units on this course: Engineering Environment Awareness, Engineering Techniques and Engineering Principles. Within each unit, the learning outcomes are covered in detail and the book includes activities and 'Test your knowledge' sections to check your understanding. At the end of each chapter is a checklist to make sure you have achieved each objective before you move on to the next section. Online, you can download answers to selected questions found within the book, as well as reference material and resources to support several other EAL units.

This book is a must-have for all learners studying for their EAL Level 2 Diploma award in Engineering and Technology and contains all of the essential knowledge you need to complete this course.

Mike Tooley has over 30 years' experience of teaching engineering, electronics and avionics to engineers and technicians, previously as Head of the Department of Engineering and Vice Principal at Brooklands College. Mike currently works as a consultant and freelance technical author.

T0141249

Engineering
Technologies   Level 2

Mike Tooley

Routledge
Taylor & Francis Group

LONDON AND NEW YORK

First published 2017
by Routledge
2 Park Square, Milton Park, Abingdon, Oxon OX14 4RN

and by Routledge
711 Third Avenue, New York, NY 10017

Routledge is an imprint of the Taylor & Francis Group, an informa business

Notice
Knowledge and best practice in this field are constantly changing. As new research and experience broaden our understanding, changes in research methods, professional practices, or medical treatment may become necessary.

Practitioners and researchers must always rely on their own experience and knowledge in evaluating and using any information, methods, compounds, or experiments described herein. In using such information or methods they should be mindful of their own safety and the safety of others, including parties for whom they have a professional responsibility.

To the fullest extent of the law, neither the Publisher nor the authors, contributors, or editors, assume any liability for any injury and/or damage to persons or property as a matter of products liability, negligence or otherwise, or from any use or operation of any methods, products, instructions, or ideas contained in the material herein.

British Library Cataloguing in Publication Data
A catalogue record for this book is available from the British Library

Library of Congress Cataloging in Publication Data
Names: Tooley, Michael H., author.
Title: Engineering technologies. Level 2.
Description: Abingdon, Oxon ; New York, NY : Routledge,
2017. | Written by Mike Tooley. | Includes index.
Identifiers: LCCN 2016013486| ISBN 9781138674479 (pbk. : alk. paper) | ISBN
9781315561295 (ebook)
Subjects: LCSH: Engineering.
Classification: LCC TA147 .T6595 2017 | DDC 620--dc23
LC record available at https://lccn.loc.gov/2016013486

ISBN: 978-1-138-67447-9 (pbk)
ISBN: 978-1-315-56129-5 (ebk)

Typeset in Helvetica by
Servis Filmsetting Ltd, Stockport, Cheshire
Printed by
Bell & Bain Ltd, Glasgow

Contents

Preface

Welcome to the challenging and exciting world of engineering! This book is designed to help you succeed on a course leading to the EAL Level 2 Diploma award in Engineering and Technology. It contains all of the essential underpinning knowledge required of a student who may never have studied engineering before and who wishes to explore the subject for the first time.

About you

Have you got what it takes to be an engineer? The EAL Level 2 Diploma in Engineering and Technology will help you find out and still keep your options open. The Diploma is often used as the technical element of the foundation modern apprenticeship framework. Successful completion of the course will provide you with a route into studying engineering at Level 3, for example, an EAL Level 3 Certificate or Diploma, or a BTEC National Diploma in Engineering, or an NVQ award studied as part of an Engineering apprenticeship.

Engineering is an immensely diverse field but, to put it simply, engineering, in whatever area that you choose, is about thinking *and* doing. The 'thinking' that an engineer does is both logical and systematic. The 'doing' that an engineer does can be anything from building a bridge to testing a space vehicle. In either case, the essential engineering skills are the same. You do not need to have studied engineering before starting a Level 2 programme. All that is required to successfully complete the course is an enquiring mind, an interest in engineering, and the ability to explore new ideas in a systematic way. You also need to be able to express your ideas and communicate these in a clear and logical way to other people.

As you study your EAL Level 2 course in Engineering and Technology you will be learning in a practical environment as well as in a classroom. This will help you to put into practice the things that you learn in a formal class situation. You will also discover that engineering is fun – it's not just about learning a whole lot of meaningless facts and figures!

How to use this book

This book provides full coverage of the three mandatory units of the EAL Level 2 Diploma in Engineering and Technology. The three mandatory units are: Engineering environment awareness, Engineering techniques, and Engineering mathematics and science principles. Within the book, each chapter is devoted to a major 'sub-group' topic. The book includes text, illustrations, examples and activities (where relevant). Each chapter concludes with a set of review questions. Answers to the review questions can be downloaded from the author's website: www.key2engtech.com.

'Test your knowledge' questions are interspersed with the text throughout the book. These questions allow you to check your understanding of the preceding text. They also provide you with an opportunity to reflect on what you have learned and consolidate this in manageable chunks. Numerical answers to the 'Test your knowledge' questions in Engineering mathematics and science principles (Unit 3) are given in Appendix 4.

Most 'Test your knowledge' questions can be answered in only a few minutes and the necessary information can be gleaned from the surrounding text. Activities, on the other hand, require a significantly greater amount of time to complete and are designed to be completed outside the classroom, often in a workshop or other practical environment. They may also require additional library or resource area research coupled with access to computing and other ICT resources. Don't expect to complete *all* of the activities in this book – your teacher or lecturer will ensure that those activities that you do undertake relate to the resources available to you and that they can be completed within the timescale of the course. Activities make excellent vehicles for improving your skills and for gathering the evidence that can be used to demonstrate that you are competent in a range of core engineering skills.

The Review questions presented at the end of each chapter are designed to provide you with an opportunity to test your understanding of each unit. These questions can be used for revision or as a means of generating a checklist of topics with which you need to be familiar. Here again, your tutor may suggest that you answer specific questions that relate to the context in which you are studying the course.

The book ends with some useful information and data presented in the form of four appendices. These include abbreviations for common terms used in engineering, information on how to use a scientific calculator, metric and imperial conversion tables, and answers to the numerical 'Test your knowledge' questions in Unit 3. To provide you with an indication of the standard that you need to reach, Appendix 1 provides you with a representative set of multiple-choice questions. More resources are available at the author's website: www.key2engtech.com.

Finally, here are a few general points worth keeping in mind:

- Allow regular time for reading – get into the habit of setting aside an hour, or two, at the weekend to take a second look at the topics that you have covered during the week.
- Make notes and file these away neatly for future reference – lists of facts, definitions and formulae are particularly useful for revision!
- Look out for the interrelationship between units and topics – you will find many ideas and a number of themes that crop up in different places and in different units. These can often help to reinforce your understanding.
- Don't be afraid to put your new ideas into practice. Remember that engineering is about thinking *and* doing – so get out there and *do* it!

Good luck with your EAL Level 2 Diploma in Engineering and Technology!

Mike Tooley

Engineering environmental awareness

CHAPTER **1**

Health and safety legislation and regulations

Learning outcomes

When you have completed this chapter you should understand the requirements of an engineering organization in meeting health and safety legislation and regulations, including being able to:

1.1 Move a load using the correct manual handling procedure.

1.2 Identify how current legislation affects the health and safety of employers, employees and the public.

1.3 State the principal provisions of the Health and Safety at Work Act.

1.4 Identify the general safe working practices associated with the operations within an engineering environment.

1.5 Identify the mandatory procedures applicable to the reporting of accidents or injuries within an engineering working environment.

Figure 1.1 Correct technique for lifting a heavy or bulky object.

Figure 1.2 Incorrect technique for lifting a heavy or bulky object.

Chapter summary

The ability to work safely in an engineering environment is essential not only for your own comfort and safety but also for the safety of those around you. This chapter will introduce you to the legislation and safety regulations that govern working practices in engineering. As you work through the chapter you will find a number of hands-on activities that will help you appreciate some of the potential hazards that exist in the workplace and the ways in which they can be minimized. Importantly, each of these activities will take you out of the classroom, preparing you for work in a real engineering environment.

Learning outcome 1.1

Move a load using the correct manual handling procedure

Working in any branch of engineering will, at some point, involve almost every engineer in performing manual tasks, such as moving or lifting equipment, tools and machines. In order to avoid accidents and injuries engineers need to adopt safe working practices and, in many cases, this will involve training in manual handling techniques. For example, lifting a heavy and/or bulky object is a task that most engineers will perform from time to time. When doing this it is important to observe the correct lifting technique which can be instrumental in avoiding back and other muscle injury. The technique is as follows:

a) Position yourself with feet slightly apart facing the object to be lifted and at a comfortable distance from it (about half an arm's length is usually ideal).

b) Bend your legs and grasp the object at each end with a firm grip (use handles whenever provided).

c) Keeping your back straight, unbend your legs slowly and evenly, raising the object to the desired height.

Figures 1.1 and 1.2 illustrate the correct and incorrect technique. When placing a heavy load on a table, bench or other work surface it is essential to ensure that the object can be adequately supported and that it will not slip or fall.

With some objects (e.g. those that are heavy, slippery or have sharp edges) it is essential to use protective clothing such as overalls, gloves and safety boots or shoes. In addition, hard hats are obligatory wherever overhead work is being carried out and when work is underground or on a construction site.

Test your knowledge 1.1

What part of the body is most likely to be injured using the incorrect lifting technique shown in Figure 1.2 and why does the technique shown in Figure 1.1 minimize this risk?

Activity 1.1

Make an eye-catching poster that can be displayed in your workshop, explaining the importance of using the correct manual handling technique.

Learning outcome 1.2

Identify how current legislation affects the health and safety of employers, employees and the public

Employees and their employers need to be fully aware of the need for any organization to meet the relevant health and safety legislation and regulations. So, before you get started on developing your engineering skills, it is essential to have an understanding of the statutory regulations and safety rules. Later you will put this knowledge to good use as you begin to practise your skills and experience some real engineering activities.

In the UK the most important health and safety legislation is:

- The Management of Health and Safety at Work Regulations 1999
- The Workplace (Health, Safety and Welfare) Regulations 1992
- The Health and Safety (Display Screen Equipment) Regulations 1992
- The Personal Protective Equipment at Work Regulations 1992
- The Manual Handling Operations Regulations 1992
- The Provision and Use of Work Equipment Regulations 1998
- The Reporting of Injuries, Diseases and Dangerous Occurrences Regulations 1995
- The Working Time Regulations 1998
- The Control of Substances Hazardous to Health Regulations 2002.

There's a lot to take in here so we shall just briefly describe the key aspects of each of these regulations:

The Management of Health and Safety at Work Regulations 1999

The duties of employers under these regulations include making 'assessments of risk' that might affect the health and safety of employees and then to act upon the risks they identify. The regulations also require employers to have an effective health and safety policy, to appoint one or more competent persons to oversee health and safety in the workplace, and to provide employees with relevant information and training.

The Workplace (Health, Safety and Welfare) Regulations 1992

These regulations require employers to provide and maintain adequate lighting, heating and ventilation in the workplace. Employers must also provide appropriate staff facilities, including toilets and areas for washing and refreshment.

The Health and Safety (Display Screen Equipment) Regulations 1992

These regulations relate to employees who regularly use display screens as part of the normal daily work. Employers are required to carry out a risk assessment of workstations and reduce any risks that they identify. Employers are also required to ensure that users take regular breaks and have regular eyesight tests. There is also a requirement to make furniture (such as workstation desks and chairs) adjustable. Display screen users

Figure 1.3 Under the Display Equipment Regulations employers are required to carry out risk assessments of CAD workstations.

must also be provided with information on how to recognize and avoid repetitive strain injury (RSI).

The Personal Protective Equipment at Work Regulations 1992

Employers must ensure that suitable personal protective equipment (PPE) is provided free of charge wherever there are risks to health and safety that cannot be adequately controlled in other ways. The PPE must be appropriate and should include items such as protective face masks and goggles, safety helmets, gloves, air filters, ear defenders, overalls and protective footwear, where required. Employers are also required to provide information, training and instruction on the use of this equipment.

Figure 1.4 Under the Personal Protective Equipment at Work Regulations employers are required to provide appropriate personal protective equipment such as hard hats, gloves and safety glasses.

The Manual Handling Operations Regulations 1992

These regulations require employers to avoid (so far as is reasonably practicable) the need for employees to undertake any manual handling activities that might involve risk of injury. The regulations require employers to make assessments of manual handling risks so that such risks are reduced. Assessments should consider the task, the load and the individual's personal characteristics (physical strength etc.). Employers should also provide employees with information on the weight of any load that they are expected to lift or move.

The Provision and Use of Work Equipment Regulations 1998

Employers are required to ensure the safety and suitability (i.e. *fitness for purpose*) of work equipment, such as machines and tools. Employers should also ensure that the equipment is properly maintained (regardless of how old it is) and provide appropriate information, instruction and training on its use. There is also a need to ensure that employees are protected from dangerous parts of machinery by the fitting of gates, barriers, guards and shields.

Figure 1.5 Under the Provision and Use of Work Equipment Regulations employers must ensure the safety and suitability of equipment such as machine tools.

The Reporting of Injuries, Diseases and Dangerous Occurrences Regulations 1995

Under these regulations, employers are required to report a wide range of work-related incidents, injuries and diseases to the Health and Safety Executive (HSE), or to the nearest local authority environmental health department. The regulations require an employer to record in an accident book the date and time of the incident, details of the person(s) affected, the nature of their injury or condition, their occupation, the place where the event occurred and a brief note on what happened. You will find more information on this in Section 1.5.

The Working Time Regulations 1998

Fatigue and tiredness not only affect judgement but are often contributing factors to workplace accidents and near-miss situations. The Working Time Regulations implement two European Community directives on the organization of working time and the employment of workers under eighteen years of age. The regulations cover the right to annual leave and to have rest breaks, and they limit the length of the working week. Importantly, employers have a contractual obligation not to require an employee to work more than an average 48-hour week (unless the worker has opted out of this voluntarily and in writing).

The Control of Substances Hazardous to Health Regulations 2002

Some engineering processes involve materials and substances that can potentially cause harm. The Control of Substances Hazardous to Health (COSHH) regulations apply to the identification, marking, handling, storage, use and disposal of such substances. Employers must provide appropriate personal protective equipment (PPE) and training in its use. First aid and emergency equipment must be made available and any harmful waste products must be disposed of safely and with consideration for the environment. Information on the storage, use, handling and disposal of hazardous substances is usually provided in the form of safety data sheets. These must be made available for reference in the workplace.

Figure 1.6 Under the Control of Substances Hazardous to Health Regulations employers must ensure that hazardous materials are labelled and stored securely.

Key point

Information on the storage, use, handling and disposal of hazardous substances is usually provided in the form of safety data sheets.

Key point

Employers must ensure that suitable personal protective equipment (PPE) is provided free of charge wherever there are risks to health and safety that cannot be adequately controlled in other ways.

Key point

Fatigue and tiredness not only affect judgement but are often contributing factors to workplace accidents and near-miss situations.

Key point

Employers are required to ensure the fitness for purpose of equipment such as tools and machines. Employers must also ensure that the equipment is properly maintained and provide appropriate information, instruction and training on its use.

Test your knowledge 1.2

State the name of the health and safety regulations that relate to:

a) the use of display screens in a computer-aided design (CAD) area
b) the maximum number of hours you can work without taking a break
c) the provision of safety guards around a power guillotine
d) the provision of washroom facilities in a workshop

Activity 1.2

Visit the COSHH section of the UK Government's Health and Safety Executive website (www.hse.gov.uk/coshh). View or download a copy of *Working with Substances Hazardous to Health: A Brief Guide to COSHH* and use it to produce an A4 'hazard checklist' that can be given to apprentices and new employees in an engineering company. Also include a list of **seven** measures that can be put in place to control the use of hazardous substances.

Learning outcome 1.3

State the principal provisions of the Health and Safety at Work Act

All work activities are covered by the Health and Safety at Work Act 1974 (HASAWA). The Act seeks to promote greater personal involvement coupled with the emphasis on individual responsibility and accountability. You need to be aware that the Health and Safety at Work Act applies to *people*, not to premises. The Act covers all employees in all employment situations. The precise nature of the work is not relevant, neither is its location. The Act also requires employers to take account of the fact that other persons, not just those that are directly employed, may be affected by work activities. The Act also places certain obligations on those who manufacture, design, import or supply articles or materials for use at work to ensure that these can be used safely and do not constitute a risk to health.

Duty of the employer

Under the Act, it is the duty of the employer to ensure, so far as is reasonably practicable, the health, safety and welfare at work of all the employees. The employer also needs to ensure that all plant and systems are maintained in a manner so that they are safe and without risk to health. The employer is also responsible for:

- the absence of risks in the handling, storage and transport of articles and substances
- instruction, training and supervision to ensure health and safety at work
- the maintenance of the workplace and its environment to be safe and without risk to health
- to provide where appropriate a statement of general policy with respect to health and safety and to provide where appropriate, arrangements for safety representatives and safety committees
- conduct his or her undertakings in such a way so as to ensure that members of the public (i.e. those not in his or her employment) are not affected or exposed to risks to their health or safety
- give information about such aspects of the way in which he conducts his undertakings to persons who are not his employees as might affect their health and safety
- in addition to having responsibilities for employees, the employer also has responsibilities to persons such as the general public (including clients, customers, visitors to engineering facilities and passers-by).

Duty of the employee

Under the Act, it is the duty of every employee while at work, to take all reasonable care for the health and safety of himself and other persons who may be affected by his acts and omissions. Employees are required to:

- cooperate with the employer to enable the duties placed on him (the employer) to be performed
- have regard of any duty or requirement imposed upon his employer or any other person under any of the statutory provisions
- not interfere with or misuse anything provided in the interests of health, safety or welfare in the pursuance of any of the relevant statutory provisions.

Shared responsibility

Importantly, it is the duty of *each person* who has control of premises, or has access to or from any plant or substance in such premises, to take all reasonable measures to ensure that they are safe and without risk. This applies equally to employers and employees and both have a hand in ensuring

Key point

Employees have general health and safety duties to:

- follow appropriate systems of work laid down for their safety
- make proper use of equipment provided for their safety
- cooperate with their employer on health and safety matters
- inform the employer if they identify hazardous handling activities
- take care to ensure that their activities don't put others at risk.

that risks and hazards in the workplace are kept to an absolute minimum.

Figure 1.7 shows typical safety notices (together with relevant safety procedures and documents) prominently displayed in an engineering workshop. Figure 1.8 shows typical markings for a fire zone and first-aid station while Figure 1.9 shows a warning notice displayed at a construction site.

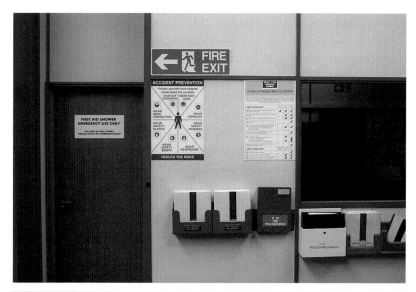

Figure 1.7 Safety notices and information prominently displayed in an engineering workshop.

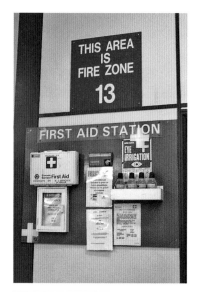

Figure 1.8 Typical fire zone and first-aid station markings.

Figure 1.9 Warning notice displayed at a construction site.

Test your knowledge 1.3

List **three** duties of a) an employer and b) an employee under the Health and Safety at Work Act.

Activity 1.3

Obtain a copy of *Reporting Accidents and Incidents at Work: A brief guide to the Reporting of Injuries, Diseases and Dangerous Occurrences Regulations 2013* (available online from the Health and Safety Executive website at www.hse.gov.uk) and use it to answer the following questions:

1 What is RIDDOR?
2 What does RIDDOR apply to?
3 In what circumstances should a report be made and who should make it?
4 Give **four** examples of major injuries that are reportable under RIDDOR.
5 Give **four** examples of dangerous occurrences that are reportable under RIDDOR.

Present your answers in the form of a handout that can be used as part of an A4 'fact sheet' for engineers and managers in an engineering company.

Learning outcome 1.4

Identify the general safe working practices associated with the operations within an engineering environment

Safe working practices are essential for anyone working in an engineering environment where a variety of different hazards are present. In this section we will examine a number of ways in which the risk of accident and injury can be minimized.

Production and manufacturing processes

Many engineering processes are potentially hazardous and these include activities such as casting, cutting, soldering, welding etc. In addition, some processes involve the use of hazardous materials and chemicals. Furthermore, even the most basic and straightforward activities can potentially be dangerous if carried out using inappropriate tools, materials and methods.

In all cases, the correct tools and protective equipment should be used and proper training should be provided. In addition, safety warnings and notices should be prominently placed in the workplace and access to areas where hazardous processes take place should be restricted and carefully controlled so that only appropriately trained personnel can be present. In addition, the storage of hazardous materials (chemicals, radioactive substances etc.) requires special consideration and effective access control.

Processes that are particularly hazardous include:

- casting, forging and grinding
- welding and brazing
- chemical etching
- heat treatment
- use of compressed air.

Figure 1.10 Welding is a hazardous process that requires special protective equipment and training.

Operation and maintenance

Hazards are associated with the operation and maintenance of many engineering systems. These include those that use electrical supplies, compressed air, fluids, gas, or petrochemical fuels as energy sources. They also include those systems that are not in themselves particularly hazardous but which are used in hazardous environments (e.g. in mining or in the oil and gas industries). Operations such as refuelling a vehicle can be potentially dangerous in the presence of naked flames or if electrical equipment is used nearby. A static discharge, for example, can be sufficient in the presence of petroleum vapour, to cause an explosion. Similar considerations apply to the use of mobile phones or transmitting apparatus in the vicinity of flammable liquids.

Electricity and electrical equipment

It should go without saying that, with the exception of portable low-voltage equipment (such as hand-held electronic test equipment), all electrical equipment should be considered potentially dangerous. This applies to all mains-operated equipment. There are three basic hazards associated with electricity:

- electric shock
- explosion due to the presence of flammable vapours
- fire due to overheating of cables and appliances.

Housekeeping and tidiness

A sign of a good worker is a clean and tidy working area. Only the minimum of tools for the job should be laid out at any one time. These tools should be organized in a tidy and logical manner so that they immediately fall to hand. Tools not immediately required should be cleaned and properly stored away (see Figure 1.11). All hand tools should be regularly checked and kept in good condition. Spillages, either on the workbench or on the floor, should always be cleaned up immediately.

Figure 1.11 A well-organized tool store where every tool can be identified.

Figure 1.12 A tidy, well-lit and well-organized workshop is a safe place to work.

Hazardous materials and processes

You are probably already well aware that many engineering processes, such as welding, casting, forging and grinding, can be potentially dangerous. You will probably also be aware that some of the materials used in engineering can be dangerous. These materials include fuels and fluids such as isopropyl alcohol and ferric chloride.

What you might not be aware of is that many processes and materials that are usually thought of as being safe can become dangerous as a result of misuse or mishandling. For example, soldering is generally considered to be a safe process; however, the fumes produced from molten solder can be highly toxic. Here, the combination of the process (soldering) with the material (flux) can result in a hazardous condition (the generation of toxic fumes).

Human carelessness

Most accidents are not caused by equipment failure but by human carelessness and negligence. This can range from 'couldn't care less' and 'macho attitudes', to the deliberate disregard of safety regulations and codes of practice. Carelessness can also result from tiredness, fatigue and ill-health and these, in turn, can result from a poor working environment.

Personal habits

Personal habits such as alcohol and drug abuse can render workers a hazard not only to themselves but also to other workers. Fatigue due to a second job (or 'moonlighting') can also be a considerable hazard, particularly when operating machines. Smoking in prohibited areas where flammable substances are used and stored can cause fatal accidents involving explosions and fire.

Supervision and training

Another contributory factor can be inadequate training. Lack of supervision can also lead to accidents when it leads to safety procedures being disregarded.

Environment

Unguarded and badly maintained plant and equipment are obvious causes of injury. However, the most common causes of accidents are falls on slippery floors, poorly maintained stairways, scaffolding and obstructed passageways in overcrowded workshops. Noise, bad lighting and inadequate ventilation can lead to fatigue, ill-health and carelessness.

Making the workplace safe

The workplace should be tidy with clearly defined passageways. It should be well lit and ventilated. It should have a well-maintained non-slip flooring. Noise should be kept down to acceptable levels. Hazardous processes should be replaced with less dangerous and more environmentally acceptable alternatives. For example, asbestos clutch and brake linings should be replaced with safer materials.

Slips and trips

Slips and trips can occur when floors and passageways are not maintained to a satisfactory standard. Slips can occur on wet floors and those contaminated with fluids such as oil and machine lubricant. Spillages must always be dealt with quickly and any area that is unsafe should be clearly marked. Trips can occur when cables and hoses are left across gangways. Cables and hoses should always be run tidily and clearly marked.

Guards

Rotating machinery, drive belts and rotating cutters must be securely fenced to prevent accidental contact. Some machines have interlocked guards. These are guards coupled to the machine drive in such a way that the machine cannot be operated when the guard is open for loading and unloading the work. All guards must be set, checked and maintained by qualified and certificated staff. They must never be removed or tampered with by operators.

Maintenance

Machines and equipment must be regularly serviced and maintained by appropriately trained and experienced personnel. This not only reduces the chance of a major breakdown leading to loss of production, it lessens the chance of a major accident caused by a plant failure. Equally important is attention to such details as regularly checking the stocking and location of first-aid cabinets and regularly checking both the condition and location of fire extinguishers. All these checks must be logged.

Safety and warning signs

Correct warning and prohibition signs should always be prominently displayed. The five main types of sign, shown in Figure 1.13, are as follows:

* prohibition signs (things that you *must not* do, e.g. 'No Smoking')
* warning signs (signs that warn you about something that is dangerous, e.g. 'Danger: High Voltage')
* mandatory signs (signs that indicate things that you *must* do, e.g. 'Eye protection must be used')

Figure 1.13 The five main types of sign.

- safe condition signs (signs that give you information about the safest way to go, e.g. 'Fire Exit')
- fire signs (signs that indicate the location of fire-fighting equipment, e.g. 'Fire Point').

Note that different colours are used to make it easy to distinguish the types of sign. For example, safe condition signs use white text with a green background, mandatory signs use white text with a blue background, and so on. It is essential that you familiarize yourself with the different types of sign and what they mean!

Personal protective equipment (PPE)

Appropriate clothing should be worn for all engineering activities. In a workshop environment, overalls or protective coats should be neatly buttoned and sleeves should be tightly rolled. Protective clothing should be sufficiently loose in order to allow easy body movement, but not so loose that it interferes with engineering tasks and activities.

As mentioned earlier, some processes and working conditions demand even greater protection, such as hard hats, ear protection, respirators and eye protection worn singly or in combination. Appropriate protective clothing must be provided by the employer when a process demands its use. It is essential to note that, by law, employers must provide this equipment and employees must make use of such equipment. Some examples are shown in Figure 1.14.

Figure 1.14 Examples of PPE.

Hazards and risks

You will need to be fully aware of the hazards and health risks associated with handling and processing engineering materials. Some materials that might at first appear to be harmless can be hazardous if they exist in certain forms (e.g. as fine airborne particles that can readily be inhaled). Some of the most common hazards include the following:

- Exposure to fibres (either glass or carbon) may cause skin rashes (occupational dermatitis) as well as irritation to the eyes, nose and throat.
- Contact with adhesives and resins may cause skin sensitization.
- Resin fumes and solvent vapours may cause irritation to the eye and nose.
- Sanding dust can be an irritant if it is inhaled.
- Protective coatings, paints and solvents may produce vapours that cause irritation to the eyes, nose, throat and lungs.
- Etching solutions and acids can cause skin irritation and burns.
- Powdered material (and dust) may cause explosions and an increased risk of static discharge.
- Use of welding, brazing and soldering equipment may result in the production of toxic fumes.
- Use of welding, brazing, hot gas and soldering equipment may cause burns.
- Incorrect use of tools (particularly machine tools, drills, lathes etc.) may cause injury to operators and other personnel.

Training

Engineering companies need to have effective procedures and training in place in order to minimize these hazards. For example, the use of a grinding wheel (see Figure 1.15) should be restricted to personnel who have received appropriate training, and appropriate eye, hand and body protection should be provided. Furthermore, the relevant health and safety advisory notices, together with any legal requirements, must be prominently displayed. Figure 1.15 shows an example of good practice.

Figure 1.15 A grinder with appropriate warning and prohibition signs. Notice also a copy of the Abrasive Wheel Regulations has been prominently displayed.

Activity 1.4

Visit your engineering workshop and carry out a detailed survey of the warning signs and notices that are displayed. Make a sketch plan of the workshop area and mark on the location of each sign or notice. Classify the signs as prohibition signs, mandatory signs, warning signs, safe condition signs and fire signs. Also find out who is responsible for placing and maintaining the workshop signs and notices.

Test your knowledge 1.4

List **three** examples of hazardous engineering processes

Test your knowledge 1.5

Identify the PPE shown in Figure 1.16.

Test your knowledge 1.6

Figure 1.17 shows a power guillotine suitable for cutting sheet metal. Identify the three signs displayed on the front of the machine and state the purpose of each sign. How is the guillotine stopped in an emergency?

Figure 1.16 See Test your knowledge 1.5.

Figure 1.17 See Test your knowledge 1.6.

Test your knowledge 1.7

Identify each of the warning signs shown in Figure 1.18.

(a)

(b)

(c)

(d)

(e)

(f)

Figure 1.18 See Test your knowledge 1.7.

Activity 1.5

Visit your engineering workshop and identify **three** items of personal protective equipment (PPE). Explain what each is used for, where it is stored and how it is used.

Fire prevention

Fire is a particular hazard in an engineering environment. Fire is the rapid oxidation (burning) of flammable materials. For a fire to start, the following are required:

- a supply of flammable materials
- a supply of air (oxygen)
- a heat source.

Once a fire has started, the removal of one or more of the above will result in the fire going out. Fire prevention is largely a matter of 'good housekeeping'. The workplace should be kept clean and tidy. Rubbish should not be allowed to accumulate in passages and disused storerooms. Oily rags and waste materials should be put in metal bins fitted with airtight lids. Plant, machinery and heating equipment should be regularly inspected, as should fire alarm and smoke detector systems. You should also know how and where to raise the fire alarm.

Smoking and naked flames must be banned wherever flammable substances are used or stored. The advice of the fire prevention officer of the local brigade should be sought before flammable substances, bottled gases, cylinders of compressed gases, solvents and other flammable substances are brought on site.

Fire procedure

In the event of you discovering a fire, you should:

- Raise the alarm and make sure that the fire service has been called.
- Evacuate the premises. Regular fire drills must be held. Personnel must be familiar with normal and alternative escape routes. There must be assembly points and a roll call of personnel.
- A designated person must be allocated to each department or floor to ensure that evacuation is complete and there must be a central reporting point so that staff can be accounted for in the event of having to evacuate a building
- Keep fire doors closed to prevent the spread of smoke. Smoke is the biggest cause of panic and accidents, particularly on

staircases. Emergency exits must be kept unlocked and free from obstruction whenever the premises are in use.

- Lifts must not be used in the event of fire.
- Only attempt to contain the fire until the professional brigade arrives if there is no danger to yourself or others. Always make sure you have an unrestricted means of escape. Saving lives is always more important than saving property.

The order in which you perform these tasks will depend on the individual circumstances. If a fire point is nearby you should raise the alarm immediately. If, however, you have to leave the room in order to raise the alarm you should close doors and windows in order to prevent the fire spreading before you exit and sound the alarm. In all cases you should alert other people to the emergency at the earliest possible stage.

Fire extinguishers

In order to cope with different types of fire several different forms of fire extinguisher are usually provided:

1 Class A extinguishers are for ordinary combustible materials such as wood, paper, cardboard and most plastic materials. This type of extinguisher is based on water and may involve the use of a wall-mounted reel and hose.
2 Class B fires involve flammable or combustible liquids such as fuels, solvents, oil and grease.
3 Class C fires involve electrical equipment, wiring, switchgear, circuit breakers and outlets (water-based extinguishers should NEVER be used on this type of fire).
4 Class D fire extinguishers are suitable for use with chemical fires involving combustible metals such as magnesium, titanium, potassium and sodium.

It is essential to remember that, while water can be an effective extinguishing agent for Class A fires (paper, wood etc.), water and air-pressurized water (APW) extinguishers **must not** be used on Class B or Class D fires because their use can actually cause the flames to spread and make the fire bigger!

Dry powder extinguishers come in a variety of types and are suitable for a combination of class A, B and C fires. These are filled with foam or powder and pressurized with nitrogen. Carbon dioxide (CO_2) extinguishers are designed for use with Class B and C fires. CO_2 extinguishers contain carbon dioxide, a non-flammable gas, and are highly pressurized. It is worth remembering that, unlike dry chemical powder extinguishers, CO_2 extinguishers don't leave a harmful residue. This can be important when dealing

Figure 1.19 A clearly marked fire exit.

Figure 1.20 A typical fire point with CO_2 and dry powder extinguishers.

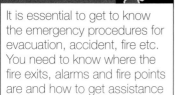

Key point

Fire can spread quickly and since it can be life-threatening knowing what action to take in order to deal with it can save lives.

Key point

It is essential to get to know the emergency procedures for evacuation, accident, fire etc. You need to know where the fire exits, alarms and fire points are and how to get assistance from a first aider or fire warden.

with fires on expensive electrical and electronic equipment which may suffer permanent damage when dry powder extinguishers are used. Finally, it is important to have appropriate training in the operation of fire extinguishers and other fire protection equipment.

Test your knowledge 1.8

What three factors need to be present in order for a fire to start?

Test your knowledge 1.9

What type of fire extinguisher should be used with a fire involving a solvent? Explain your answer.

Learning outcome 1.5

Identify the mandatory procedures applicable to the reporting of accidents or injuries within an engineering working environment

Earlier in Section 1.1 we mentioned the Reporting of Injuries, Diseases and Dangerous Occurrences Regulations (RIDDOR). RIDDOR is the law that requires employers, and other people in control of work premises, to report and keep records of:

- work-related accidents which cause death or result in serious injuries
- diagnosed cases of certain industrial diseases
- dangerous occurrences with the potential to cause harm.

Reporting certain incidents is a legal requirement. The report is designed to inform the appropriate enforcing authorities (such as the Health and Safety Executive [HSE], local authorities and the Office for Rail Regulation) about deaths, injuries, occupational diseases and dangerous occurrences, so they can identify where and how risks arise, and whether they need to be investigated. This helps the enforcing authorities to target their work and provide advice about how to avoid work-related deaths, injuries, ill-health and accidental loss.

For the purposes of RIDDOR, an accident is a separate, identifiable, unintended incident that results in physical injury. Not all accidents need to be reported. A RIDDOR report is required only when the accident is work-related and it results in an injury of a type that's reportable such as a fracture (other than to fingers, thumbs and toes); loss of an arm, hand, finger, thumb, leg, foot or toe; permanent loss or reduction of sight; crush injuries leading to internal organ damage; serious burns (covering more than 10% of the body, or damaging the eyes, respiratory system or other vital organs).

Injuries to non-workers (e.g. visitors or members of the public) must be reported if a person is injured and is taken from the scene of the accident to hospital for treatment as a consequence. There is no requirement to establish what hospital treatment was actually provided, and no need to report incidents where people are taken to hospital purely as a precaution when no injury is apparent.

Reportable dangerous occurrences

Reportable dangerous occurrences are certain, specified near-miss incidents with the potential to cause harm. Not all such events require reporting. There are 27 categories of dangerous occurrences that are relevant to most workplaces. For example, the collapse, overturning or failure of load-bearing parts of lifts and lifting equipment; plant or equipment coming into contact with overhead power lines; explosions or fires causing work to be stopped for more than 24 hours. Additional categories of dangerous occurrences apply to mines, quarries, offshore workplaces and railways.

Recording

A record must be kept of any accident, occupational disease or dangerous occurrence that requires reporting under RIDDOR. These records are important because they ensure that organizations collect sufficient information to allow them to properly manage health and safety risks. This information is a valuable management tool that can be used as an aid to risk assessment, helping to develop solutions to potential risks. In this way, records also help to prevent injuries and ill-health, and control costs from accidental loss.

Depending upon the circumstances, different types of report should be made to the HSE. Reports can be made using online forms covering reporting of:

* injuries
* dangerous occurrences
* injuries offshore
* dangerous occurrences offshore

Key point

RIDDOR makes it a legal requirement for organizations to report work-related accidents that cause death or result in serious injuries as well as dangerous occurrences with the potential to cause harm.

Key point

Accident records are important because they ensure that organizations collect sufficient information to allow them to properly manage health and safety risks.

- diseases
- flammable gas incidents
- dangerous gas fittings.

A typical injury report is shown in Figure 1.21.

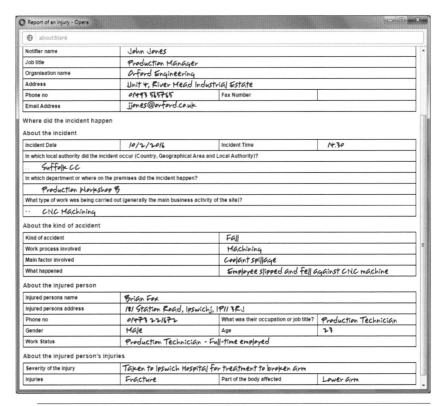

Figure 1.21 A typical injury report.

Test your knowledge 1.10

How does RIDDOR define the term 'accident'?

Test your knowledge 1.11

Give **two** examples of near-miss incidents that should be reported under RIDDOR as 'dangerous occurrences'.

Test your knowledge 1.12

Refer to the accident report shown in Figure 1.21 and answer the following questions:

1 On what date and at what time did the accident occur?
2 What was the name of the injured person?
3 What injuries were sustained?
4 What was the job role of the injured person?
5 Who made the accident report and what was his job title?

Activity 1.6

Figure 1.22 shows part of an engineering workshop. Look at the photograph carefully and identify **six** hazardous conditions. For each condition that you identify state what action would be required to put things right and minimize the hazard.

Figure 1.22 See Activity 1.6.

Review questions

1. What additional precautions might be necessary when lifting a heavy object that has sharp edges?

2. What regulations require employers to provide and maintain adequate lighting and ventilation in the workplace?

3. What regulations are designed to help reduce the incidence of repetitive strain injury (RSI) when people are regularly working with computers?

4. What regulations relate to the storage, handling and disposal of hazardous substances?

5. State **four** duties of employers under the Health and Safety at Work Act.

6. State **two** potential hazards associated with using fibrous materials.

7. Explain what is meant by a 'reportable dangerous occurrence'. Why is it necessary to report such an incident?

8. What are 'control measures' and how might they be applied in the case of fumes in a workshop used for welding?

9. Explain the procedure that should be followed when a fire is discovered.

10. State **three** types of fire extinguisher. What types of fire should be tackled with each of these?

Chapter checklist

Learning outcome	Page number
1.1 Move a load using the correct manual handling procedure.	4
1.2 Identify how current legislation affects the health and safety of employers, employees and the public.	5
1.3 State the principal provisions of the Health and Safety at Work Act.	10
1.4 Identify the general safe working practices associated with the operations within an engineering environment.	13
1.5 Identify the mandatory procedures applicable to the reporting of accidents or injuries within an engineering working environment.	24

CHAPTER 2

Roles and responsibilities

Learning outcomes

When you have completed this chapter you should understand the roles and responsibilities of both employers and employees in the continuous development of skills and working relationships. You should be able to:

2.1 State the roles and responsibilities of both the employer and employee in the development of an effective workplace.

2.2 State the roles and responsibilities of both the employer and employee in the process of personnel development.

Chapter summary

This chapter is all about the work that engineers do and, in particular, their roles and responsibilities. But, before we begin, it's very important to realize that this work can be extremely diverse. The online *Cambridge Dictionary of English* defines an engineer as 'a person whose job is to design or build machines, engines or electrical equipment, or things such as roads, railways or bridges'. Sadly, apart from telling you that he or she is involved in designing and building things, this definition doesn't give you much idea of what an engineer actually does. What should become obvious, however, is that the work of an engineer is rather important as it helps us make the best of the world around us. And, as with any other job, there are various key roles and responsibilities that relate to working in the industry.

Learning outcome 2.1

State the roles and responsibilities of both the employer and employee in the development of an effective workplace

The term 'engineer' can be applied to a wide range of different occupations, each with different roles and responsibilities. Here are just some of the diverse job roles that engineers can have:

Construction engineers

Construction engineers work in the building and construction sector where they are responsible for preparing a site, organizing the materials and resources that will be required to deliver a particular project or facility.

Design engineers

Design engineers specify the optimum production methods, materials and processes required to turn an idea into reality. They will usually develop and test several models and prototypes of an end product in order to verify its operation, making modifications and improvements along the way. They will also develop a full set of documentation for a product, including working drawings, a *bill of materials* and a detailed performance specification. In conjunction with a Production Engineer, they may also be involved with recommending and specifying materials and manufacturing methods.

Figure 2.1 Design engineers turn ideas into reality.

Development engineers

Development engineers apply the results of research in the development of new or improved products. They turn ideas into reality by effectively bridging the gap between research and manufacture.

Plant engineers

Plant engineers control machines, manufacturing plant and equipment. They may also be concerned with energy supply, power, transport and communication.

Production engineers

Production engineers are concerned with manufacturing and processing methods including plant layout, machine tool and equipment selection, and assembly methods. They need to consider production flow, energy and material supply, storage and disposal of waste. They select processes and tools, integrate the flow of materials and components, and provide a means of testing and inspecting the product.

Quality engineers

Quality engineers are responsible for ensuring that a product or process conforms to specification. Like test engineers they often

Figure 2.2 A production engineer programming a CNC milling machine.

work closely with production engineers, identifying problems and ensuring that production is of consistent quality.

Research engineers

Research engineers are concerned with the development of new processes and the application of new materials and technologies. Not only do they look for new cost- and energy-efficient ways of solving existing problems but they also explore entirely new applications for existing engineering technologies.

Sales engineers

Sales engineers work directly with customers and clients. They may be involved with preparing specifications and contracts as well as raising awareness of a company's ability to supply goods and services, matching these to customers' needs. Because they have regular face-to-face contacts with clients, sales engineers have an important liaison role. Sales engineers combine an understanding of the products and services that a company supplies with an ability to sell and attract new business.

Test engineers

Test engineers work closely with production engineers. They are often involved with testing, adjusting and/or calibrating equipment and manufactured products. They may also work as 'trouble-shooters' helping to identify problems with the production process, materials, parts and components used.

Figure 2.3 A test engineer discusses a CNC manufacturing process with a production engineer.

Engineers and the workplace

Despite the diverse job roles that we've just mentioned, engineers and their employers have a common set of roles and responsibilities that relate to the place in which they work, making it a safe and efficient environment and ensuring that relevant company policies are followed. This might typically involve:

- maintaining and updating records
- reporting issues relating to health and safety
- communicating effectively with managers and other team members
- obtaining and sharing information with other team members
- regularly updating skills and techniques
- seeking and undertaking training where necessary
- contributing to programmes that improve efficiency and cost-effectiveness
- providing effective feedback to customers, clients and others in the workplace.

Employers must also play their part in all of this by:

- providing a working environment that is safe and fit for purpose
- maintaining effective health and safety policy and procedures
- providing a means of recording work and tracking progress
- communicating effectively with personnel at all levels within the organization
- ensuring that individuals and work teams have the information that they need

Key point

Engineers and their employers have a shared set of roles and responsibilities that relate to the place in which they work, making it a safe and efficient environment and ensuring that relevant company policies are followed.

- reviewing the skills, technologies, materials and equipment needed to do the job and ensuring that these are available when required
- providing regular opportunities for training and updating skills
- supporting employees at all levels with effective training and an individually focused professional development programme.

Figure 2.4 Engineers frequently work in teams where ideas can be shared among team members.

Test your knowledge 2.1

1 Explain, briefly, the role of a) a design engineer and b) a sales engineer.

2 List **four** ways in which a) an employee and b) an employer can contribute to the development of an effective workplace.

Activity 2.1

Attend an engineering team meeting in your workplace, training centre or college. Make a note of who attends the meeting and what their job roles are. Listen to what is said and note down any decisions that are made. Also note down any agreed action points and the names of those who will be responsible for carrying out the action. Finally, don't forget to thank the chairperson for letting you attend the meeting.

Learning outcome 2.2

State the role and responsibilities of both employer and employee in the process of personnel development

Within any engineering organization people are a fundamental and vitally important asset. In fact, they are even more important than the equipment, materials, buildings and machines that a company needs for its day-to-day operation. Without people, engineering operations cannot be carried out. Because of this, investing in human resources is vital for any organization. Being able to attract staff with the right skills, knowledge, ability and motivation is just a first step. Being able to retain these staff by offering incentives such as good working conditions, competitive pay, training and development is essential. After all, there's no point in wasting time, effort and resources in the recruitment of staff if they leave after 18 months and take their skills and knowledge elsewhere. And it could be even worse if they take these skills to a competitor.

The role of a company's Human Resources department is crucial in attracting, motivating, supporting and retaining high-quality staff. Successful engineering companies recognize this fact by investing considerable sums of money in their personnel, training and development functions.

Your own personal development

At this point it is well worth thinking about your own personal development and, in particular, developing an outline plan for your own academic and career development. To do this you will need to reflect on your progress and achievements, identifying the knowledge and skills that you wish to develop and improve on. Your own personal development plan will help you to:

- identify gaps in your knowledge and skills
- locate the resources and support that you need
- make the most of opportunities that present themselves
- set realistic goals and targets for your own development
- identify training and areas for specific development
- improve your employability, preparing you for the next step on your career ladder.

Finally, you should always feel free to discuss your personal development with your supervisor, tutor or line manager. All good employers will provide you with an opportunity for formal work appraisal on a regular one-to-one basis. This can be invaluable in focusing on your needs and aspirations and it will help you to formalize your ideas with input from your supervisor, tutor or line manager.

Key point

The role of a company's Human Resources department is crucial in attracting, motivating, supporting and retaining high-quality staff.

Figure 2.5 Appraisal will provide you with a way of sharing ideas and discussing your own personal development with your supervisor or line manager.

Test your knowledge 2.2

Explain, briefly, why a company invests in its Human Resources department.

Activity 2.2

Make a list of key points that you might want to cover at an appraisal with your supervisor or training manager. What questions would you want to ask and why would you want to ask them?

Review questions

1. Briefly explain the role of a) a production engineer and b) a test engineer.

2. State **three** responsibilities of an employee that contribute to the development of an effective workplace.

3. State **three** responsibilities of an employer that contribute to the development of an effective workplace.

4. Briefly explain **three** advantages of having a personal development plan.

5. Briefly explain what is meant by an appraisal system and how it can help you with your own personal development.

Chapter checklist

Learning outcome	Page number
2.1 State the roles and responsibilities of both the employer and employee in the development of an effective workplace.	30
2.2 State the role and responsibilities of both employer and employee in the process of personnel development.	35

CHAPTER **3**

The engineering environment

Learning outcomes

When you have completed this chapter you should understand the internal and external environments associated with the operations of an engineering organization. You should be able to:

3.1 State the main department functions within an engineering organization.

3.2 Identify the types of engineering organizations in the UK.

3.3 Identify the factors that influence change within the engineering industry.

3.4 Identify the effects of industrial change on the requirements of the workforce.

3.5 Identify the different categories of employees within the engineering industry.

Chapter summary

The environment in which an engineering organization operates has a major impact on the way it works. The internal environment is known and predictable but many external factors, such as the general state of the economy, are beyond the control of a company. However, some of these can be foreseen, in which case they are predictable (and can thus be planned for). Other external factors are largely unforeseen and as a result may pose a threat to a company's profitability. For example, a recent fire in a Japanese semiconductor manufacturing plant which gave rise to a global shortage in memory chips had a significant impact on the cost and availability of certain types of computer system. Some environmental factors can be regarded as threats while others can be regarded as opportunities. For example, the availability of a new manufacturing process that significantly reduces costs can be considered to be an opportunity. Conversely, falling consumer demand resulting from high interest rates can be considered a threat. Change is a factor that has been ongoing and relentless and responding to change is perhaps the greatest challenge of all for any engineering company. Why? Because, depending upon whether or not a company is ready and able to respond to it, change can be regarded as both a threat and an opportunity!

Learning outcome 3.1

State the main department functions within an engineering organization

Depending on the diversity of its products and services an engineering organization will have several different departments, each with a different function. These usually include:

- design and development
- production and manufacture
- management
- sales and marketing
- human resources.

Engineers may be associated with several of these departments so we shall now briefly look at the way these different departments work and the work of engineers within them. A visit to the workplace can be extremely useful in helping you to understand the work role and the way in which engineering companies are organized.

Design and development

New product design and development is often a crucial factor in the survival of a company. In an industry that is fast changing, firms must continually revise their design and range of products. This is necessary because of the relentless progress of technology as well as the actions of competitors and the changing preferences of customers. This area is the province of the *design engineer*.

Production and manufacturing

Once the final design has been completed the focus passes to the next stage, i.e. that of actually manufacturing the product. This is the role of the *production department*. Completely new products will often have different characteristics, and will perhaps be made from different materials from previous products with similar functionality. They may also require different processes in their manufacture. All of this will require liaison between *design engineers* and *production engineers* on methods for production and in deciding what manufacturing equipment, machine tools and processes are required.

Figure 3.1 Different production processes may require flexible factory layouts.

Also important is the control of raw materials and component stocks, especially the levels of *work-in-process*. Finance will want to restrict stock levels to reduce the amount of capital tied up in stocks, while the production manager will be concerned with having sufficient stock to maintain production, but avoiding congestion of factory floor space. Typical reporting lines within a large engineering company are shown in Figure 3.2.

Key point

The production department is at the heart of any manufacturing business. It translates the designs for products, which are based on market analysis, into the goods wanted by customers.

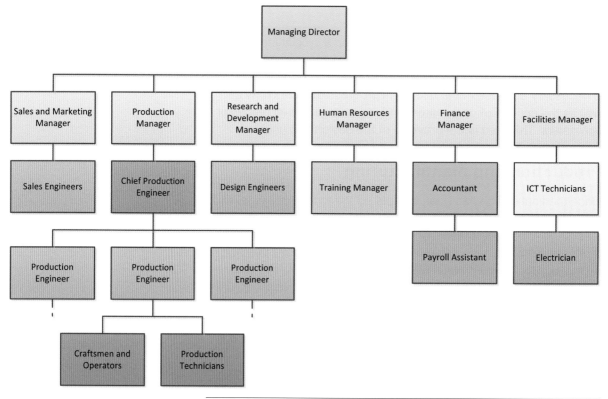

Figure 3.2 Typical reporting lines within an engineering company.

Test your knowledge 3.1

List each of the main stages in the development of a new product.

Process and facilities management

Decisions have to be made in relation to location of the production plant and the design and layout of manufacturing facilities. The design of production processes also requires interaction with design engineers and marketing functions. Selecting the process of production is important and is strategic in nature. This means that it has a wide impact on the operation of the entire business. Decisions in this area can bind the company to a particular kind of equipment and labour force because the large capital outlay that must be made will limit future options. For example, a motor manufacturer has to commit very large expenditures to lay down plant for production lines to mass-produce cars. Once in production, the company is committed to the technology and the capacity created for a long time into the future.

Production methods

There are three basic methods for production processes:

- line flow
- intermittent flow
- project-based.

Line flow is the type of system used in the motor industry for assembly lines for cars. It also includes continuous type production of the kind that exists in the chemicals and food industries. Both kinds of line flow are characterized by linear sequences of operations and continuous flows and tend to be highly automated and highly standardized.

Intermittent flow is the typical batch production or job shop, which uses general purpose equipment and highly skilled labour. This system is more flexible than line flow, but is much less efficient than line flow. It is most appropriate when a company is producing small numbers of non-standard products, perhaps to a customer's specification.

Finally *project-based production* is used for unique products which may be produced one at a time. Strictly speaking, there is not a flow of products, but instead there is sequence of operations on the product which have to be planned and controlled. This system of production is used for prototype production in R&D and is used in some engineering companies that produce major machine tool equipment for other companies to use in their factories.

Marketing

Marketing is all about matching company products with customer *needs*. If customer needs are correctly identified and understood, then products can be made which will give the customer as much as possible of what he or she wants. Companies that view the customer as key are those companies that tend to stay in business, because customers will continue to buy products that meet their requirements. Hence marketing activities are centred on the process of identifying and filling customers' known needs as well as discovering needs the customer does not yet know, and then exploiting this by finding out how to improve products so that customers will buy this company's products in preference to other goods. Some of the most important activities are:

- market research
- monitoring trends in technology and customer tastes
- tracking competitors' activities

Key point

The process of production has a wide impact on the operation of the entire business and decisions in this area can bind the company to a particular kind of equipment and workforce, because the large capital outlay that must be made will limit future options.

Key point

Marketing is about matching a company's products and services with customer needs. If these are correctly identified and understood, then products can be made which will give the customer as much as possible of what he or she wants.

- promotion activities
- preparing sales forecasts.

Test your knowledge 3.2

Distinguish between line flow and project-based production methods.

Learning outcome 3.2

Identify the types of engineering organizations in the UK

The various types of engineering organization in the UK can be categorized by the sector in which they are active. Note, however, that some large corporations are active in more than one sector. These are just some of the engineering sectors found in the UK:

- aerospace and aeronautical engineering
- agricultural engineering
- automotive engineering
- chemical engineering
- civil engineering and construction
- computer engineering and information technology
- consulting services
- domestic equipment manufacture
- electronics and telecommunications
- electricity generation, supply and utilization
- marine engineering
- mechanical engineering
- nuclear engineering
- oil and gas production and distribution
- railway engineering.

Key point

Engineering companies can be classified according to the sector(s) in which they are active.

Test your knowledge 3.3

Identify the sector (or sectors) in which the following engineering companies are currently active:

a) Jaguar Land Rover
b) Rolls-Royce

c) Siemens
d) BAE Systems
e) Dyson
f) WS Atkins

Activity 3.1

Carry out research to identify at least **three** established engineering companies that are active within 30 miles of your home. Identify the engineering sector(s) that each company is active in and list the products and services that each company provides.

Learning outcome 3.3

Identify the factors that influence change within the engineering industry

Internal and external factors

There are many factors that bring change to all engineering organizations including:

- markets and the general state of the economy
- consumer demand
- demographic and social trends
- competitive products and services
- consumer confidence and customer/client relationships
- innovation and technological change.

It's worth noting that, to some extent, all of these factors are present to a varying degree all of the time. Figure 3.3 shows factors that affect an engineering company. Some factors can be considered to be *internal* as they are within the direct control of the company. Others are considered to be *external* because the organisation has no direct control over them. Internal factors are relatively easy to influence. External factors, on the other hand, may be difficult if not impossible to influence.

Key point

Many environmental factors, both internal and external, bring change to an engineering company. Internal factors can be easy to control while external factors may be difficult or impossible to influence.

Test your knowledge 3.4

Identify three external factors that can have a major impact on all engineering organizations.

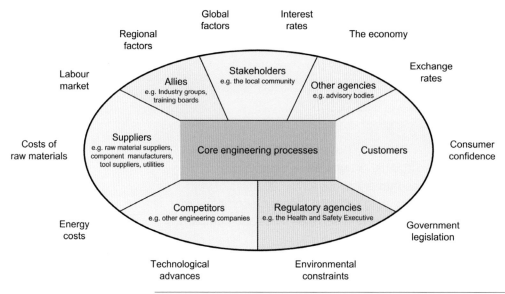

Figure 3.3 Factors that affect the operation of an engineering company.

Location

For the first half of the twentieth century, engineering manufacturing was usually located within cities. Since then there has been a tendency for any new engineering enterprise to be located in an industrial estate or business park on the periphery of a town rather than in the town itself. This is because of the following:

- The town centre is already too congested to allow for additional new industry.
- There are advantages to being located in purpose-built industrial accommodation.
- Such sites often have good road links with the national motorway network.
- Engineering activities that may involve noise and other pollutants are best kept away from the commercial and domestic centres of towns.

Regional shift

Over the last 50 years there has been a shift of engineering away from the old industrial regions such as the north-east and Midlands of England and parts of Scotland to more convenient locations such as the Thames Valley along the M4 motorway and along the M11 motorway north of London. The reasons for the regional shift are many and varied.

- The cost and pollution-causing record of coal means that it is no longer a popular fuel.

Figure 3.4 Modern industrial units are located outside towns and cities.

- With natural gas and electrical power being available almost anywhere in the country, new engineering activities can be located in regions having pleasant natural and social environments.
- The ubiquitous motor car, good roads and frequent air services mean that commuter and business communications to most regions are no longer a major problem.
- There is good availability of a pool of technologically skilled labour in places where high-technology companies are clustered together.

European Union

Within the EU, engineering activities have the usual varied pattern. The favoured countries are those which were the first to industrialize in the eighteenth and nineteenth centuries. Britain, Germany, France and Italy are predominant in Europe with the main concentration lying within a rough triangle formed by London, Hamburg and Milan. Ireland, Spain, southern Italy and Greece lie outside this triangle and tend to be less industrialized.

The past 30 years have seen a shift in some of the major engineering activities which used to be concentrated in Europe, North America and Japan. In particular, much of the electronics and printing industries have migrated to the Pacific rim countries such as Hong Kong, Singapore, Taiwan, Thailand and more recently Indonesia. The main reason for this shift is the low labour costs to be found in the Far East.

The global economy

The global economy also impacts on the operation of engineering companies based in the UK. This can be due to various factors, but during periods of global recession it can be difficult to sell goods and services abroad. It can also be difficult to sell products and services when the UK economy is strong and the value of sterling is high compared with other currencies such as the euro and the US dollar. In recent years the decline in growth of Far Eastern economies, notably Japan and China, has had the effect of flooding the European market with cheap components and materials. This has been bad news for several UK-based companies, but there is still cause for optimism as the UK has been one of the best performing European countries and one of the first to emerge from the European recession of the last decade.

New technologies

In the last 20 years there has been a major revolution in the design and manufacture of a wide variety of engineered products. For example, GPS units are now built into most modern vehicles, offering drivers a means of navigating without the need of maps and road signs. Another example is the ability to connect a wide range of audiovisual equipment to the Internet so that media can be streamed directly from the Web.

Among the most significant new technologies are:

* networked systems where distributed computers and microcontrollers work together in manufacturing and production
* the introduction of automated and robotic systems
* the use of embedded computers and microcontrollers (i.e. computers that can be integrated into products)
* the widespread availability of fast Internet connections allowing data to be exchanged on a global basis
* the availability of 'smart' materials and coatings with superior properties that make them increasingly more attractive for use in manufacturing.

Key point

New technology has had an increasing impact on the way in which engineering companies operate as well as on the products and services that they deliver.

Test your knowledge 3.5

GPS is a new technology. Explain how this technology has changed the way we live and give **three** examples of how engineering companies have embraced the technology in their products and services.

Test your knowledge 3.6

Classify each of the following as a threat or an opportunity for a small engineering company:

a) interest rates climb to an all-time high
b) the government introduces low-cost business development loans
c) a major competitor is offering high salaries to new recruits
d) a local college starts a new course in advanced manufacturing technology

Activity 3.2

For one of the three companies that you've identified in Activity 3.1, investigate how the company (and its products and services) has changed over the past 25 years. Identify factors that may have caused this change.

Learning outcome 3.4

Identify the effects of industrial change on the requirements of the workforce

The ongoing changes that have affected engineering have had a major impact on the workforce. As a consequence of the introduction of machine tools and automation there is less need

Figure 3.5 Training and development is crucial when new technologies and production methods are introduced.

for unskilled and semi-skilled manual workers. Instead, there has been an increasing requirement for engineers to be able to use and operate computer systems, for example, programming computer numerically controlled (CNC) machines. There's also been a need for engineers to develop a range of new skills including the use of analytical software, computer simulation and modelling.

Transferable skills

Due to the need to adapt to change and uncertainty, the entire UK workforce (not just the engineering workforce) will need to be more flexible, willing to accept change and acquire new skills. Fifty or more years ago many school leavers might have remained at work in their first job for all of their working lives. This is no longer the case and most of us can expect to move on several times as our career develops. Transferable skills, such as learning to use a computer to set up a spreadsheet, can be taken from job to job, from work role to work role, and from company to company. Transferable skills are applicable anywhere and time invested in acquiring and practising them is time well spent.

> **Key point**
>
> Transferable skills are applicable anywhere and time invested in acquiring and practising them is time well spent.

Figure 3.6 Computer skills, such as coding and CAD, are widely transferable.

Test your knowledge 3.7

Explain what is meant by a transferable skill. Give **two** examples of transferable skills.

The Engineering Council divides engineers into four specific categories: chartered engineers, incorporated engineers, engineering technicians and ICT technicians. These categories are not used in all branches of engineering, but the roles are generally well understood and serve as useful benchmarks with which to compare the roles of engineers working in a wide variety of engineering sectors. The roles can be briefly summarized as follows:

Chartered engineers (CEng)

Chartered engineers are characterized by their ability to develop appropriate solutions to complex engineering problems, using new or existing technologies. Engineers are variously engaged in technical and commercial leadership and must have effective interpersonal skills. The production manager and chief production engineer in Figure 3.2 would normally be chartered engineers.

Incorporated engineers (IEng)

Incorporated engineers maintain and manage applications of current and developing technology, and may undertake engineering design, development, manufacture, construction and operation. Incorporated engineers are variously engaged in technical and commercial management and possess effective interpersonal skills. The production and sales engineers in Figure 3.2 would normally hold this qualification.

Engineering technicians (EngTech)

Engineering technicians are involved in applying proven techniques and procedures to the solution of practical engineering problems. They carry supervisory or technical responsibility, and are competent to exercise creative aptitudes and skills within defined fields of technology. Engineering technicians contribute to the design, development, manufacture, commissioning, operation or maintenance of products, equipment, processes or services. The production technicians in Figure 3.2 would normally hold this qualification.

Information and Communication Technology technicians (ICTTech)

ICT technicians work in areas that involve the use and application of computer systems. Their work involves specifying, installing, configuring and maintaining hardware and software systems as well as the peripheral devices and systems that are connected to them for control, data logging and supervisory functions. Their skills are particularly in demand where automated systems are used and where computers are playing an increasingly important role in design, manufacture and production. The ICT technician in Figure 3.2 would normally hold this qualification.

Craftsmen and operators

In addition to these four roles, other skilled staff may be required to operate machine tools such as lathes, welding equipment, cranes and lifting apparatus, and also for numerous routine tasks associated with manual production and assembly. These people have specialized skills and their work is usually restricted to a particular set of repetitive tasks. They would normally hold specialized vocational qualifications.

Test your knowledge 3.8

Describe the **four** types of engineer defined by the Engineering Council.

Review questions

1. Explain what is meant by *marketing*. Why is marketing important?

2. List **four** department functions present within most engineering companies.

3. Identify **four** different types of engineering organization in the UK.

4. List **three** factors that influence change within the engineering industry.

5. Identify **four** different categories of employee within the engineering industry.

6. Explain the effect of industrial change on the engineering workforce. What new skills are likely to be required from engineers in the next decade?

7. Give **one** example of the impact of economic change on the operation of UK-based engineering companies.

8. Give **one** example of the impact of technological change on the operation of UK-based engineering companies.

Chapter checklist

Learning outcome	Page number
3.1 State the main department functions within an engineering organization.	40
3.2 Identify the types of engineering organizations in the UK.	44
3.3 Identify the factors that influence change within the engineering industry.	45
3.4 Identify the effects of industrial change on the requirements of the workforce.	49
3.5 Identify the different categories of employees within the engineering industry.	51

Engineering techniques

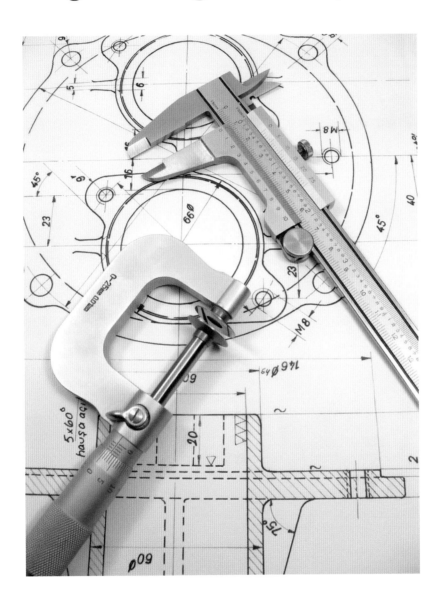

CHAPTER 4

Engineering communication

Learning outcomes

When you have completed this chapter you should understand the forms of communication used within engineering, including being able to:

4.1 Draw an orthographic drawing.

4.2 State the merits and limitations of different forms of communication:

- verbal
- written
- electronic means.

4.3 Identify basic drawing conventions, layouts and the use of sketches.

4.4 Identify types of lines, detailing and dimensioning.

Chapter summary

Effective communication skills are essential for anyone working in engineering. This chapter will help you develop skills in using and interpreting information in a wide variety of forms. It aims to provide you with experience of speaking, reading and writing as well as graphical means of communication including drawing and sketching. These skills are essential not only for employment in engineering but also as a basis for further study.

Throughout this chapter there are numerous opportunities to develop your own skills in obtaining, processing, evaluating and presenting information. In order to do this, you should be prepared to make appropriate use of IT and ICT. In the next chapter we will look at this in much greater detail.

Learning outcome 4.1

Draw an orthographic drawing

Engineers rely heavily upon sketching and drawing as a means of communication. As an engineer you must be able to read and use working drawings as well as produce them using both hand-drawn and computer techniques. To avoid confusion, your engineering drawings must comply with recommended standards and conventions. You will also need to be able to read schematic diagrams such as those used in electronics, pneumatics and hydraulics. In this section we will introduce you to a common type of engineering drawing known as an *orthographic drawing* but, before we do, it's important to realize that this is just one of several types of drawing used in engineering.

Formal and informal drawings

Engineering drawings are sometimes referred to as *formal* or *informal*. Informal engineering drawings, like the one shown in Figure 4.1, can range from quick freehand sketches to simple charts and diagrams. These are intended to convey a quick impression of what something will look like or how something will work. Formal drawings, like the one shown in Figure 4.2, generally take much longer to produce and usually contain a lot more detail. They are also much more precise and often include a scale and dimensions.

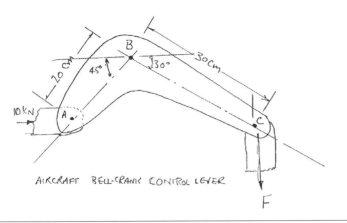

Figure 4.1 A quick sketch is an example of an informal drawing.

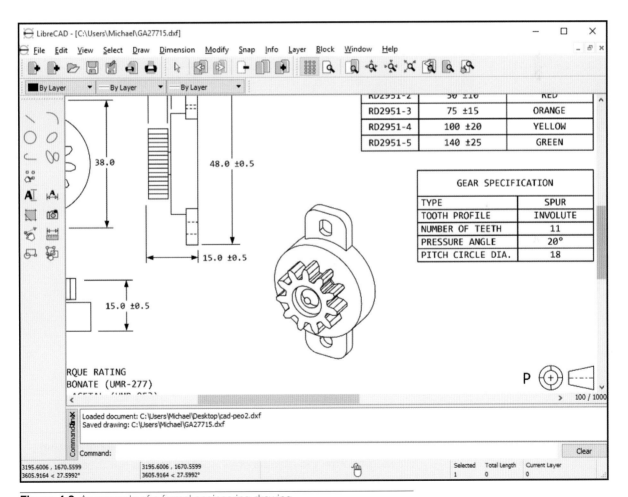

Figure 4.2 An example of a formal engineering drawing.

Sketches

Sketching is one of the most useful tools available to the engineer to express his or her ideas and preliminary designs. Sketches are drawn freehand and they are used to gain a quick impression of what something will look like. Where appropriate, labels, approximate dimensions and brief notes can be added. Engineering drawing isn't just about using sophisticated computer-aided design (CAD) workstations to produce formal drawings. Being able to produce a quick sketch is often very useful so don't be afraid to pick up a drawing pad and put your ideas into visual form!

> **Key point**
>
> Because they can convey a great deal of information very quickly, engineers make a great deal of use of drawings. Depending upon the way they are presented, drawings are often classified as either formal or informal.

> **Key point**
>
> Engineers use hand-drawn sketches to convey their ideas quickly and without having to use a lot of words. Sketches can be 2D or 3D drawings and may, or may not, include dimensions.

Activity 4.1

Produce a 3D sketch of an engineering component like that shown in Figure 4.3. Include in your sketch approximate dimensions and brief details of the materials used.

Figure 4.3 See Activity 4.1.

Producing a formal engineering drawing

Formal engineering drawings were once produced by hand using drawing boards and manual drawing instruments but these days they are invariably produced using powerful computer-aided design (CAD) software. Developments in software and desktop computers have reduced the cost of CAD and made it more powerful. At the same time, because computer-aided drawing does not require the skill required for manual drawing (which can take years of practice to achieve) CAD has become more 'user friendly' and has a number of important advantages over manual drawing, notably:

- *Accuracy.* The dimension and relative positioning of features in the drawing do not rely on the quality of the eyesight and manual dexterity of the person making the drawing. CAD offers a degree of precision that is very much greater than that of manual drawing.
- *Repetitive features.* Many drawings incorporate features that are repeated, often at constant intervals. CAD allows you to create a particular feature and then repeat it wherever needed. For example, a number of holes round a circle do not have to be individually drawn but can be easily produced automatically by rotating or mirror imaging. Similarly, lines can easily be drawn at a constant and accurately pre-determined linear or angular spacing from one another.
- *Blocks.* CAD software allows you to save part of a drawing in a library so that it can be easily re-used in the same or other drawings. This means that having created a component in your drawing you never have to create it again. Libraries can be shared between CAD users and this ensures a high degree of consistency.
- *Editing.* Every time you erase and alter a manually produced drawing on tracing paper or plastic film the surface of the drawing is increasingly damaged. Using CAD you can erase and redraw as often as you like with no ill effects. You can also make use of *Undo* and *Redo* features that allow you to backtrack and move forward with one change to the drawing at a time.

Figure 4.4 A modern CAD workstation.

- *Templates.* Commonly used drawings can be saved as templates so there is no need to start from scratch every time a new drawing is required.
- *Storage.* No matter how large and complex the drawing, it can be stored digitally and distributed electronically. Copies can be archived or shared quickly and easily without the risk of damage or loss.
- *Prints.* Hard copy, both black and white and colour, can be produced easily when and where required. Manually produced drawings must be scanned or photocopied, which reduces quality and introduces errors.

Orthographic drawing

Engineering drawings are normally produced by a technique known as *orthographic drawing*. The word orthographic means to draw at right angles and it is derived from the Greek words *orthos*, meaning straight, rectangular and upright, and *graphos* meaning written down. Orthographic drawing techniques allow us to represent three-dimensional solids on a two-dimensional surface so that all the dimensions are true length and all the surfaces are true shape. To achieve this when surfaces are inclined to the vertical or the horizontal we have to use auxiliary views, but more about these later. Let's keep things simple for the moment.

Engineers use two orthographic drawing techniques, either first-angle or third-angle projection. The former is also called *English projection* and the latter is sometimes known as *American projection*. You need to be familiar with both of these techniques.

First-angle projection

Figure 4.5a shows a simple component drawn as a 3D object (known as an *isometric view*). Figure 4.5b shows the same component as an orthographic drawing. This time we make no attempt to represent the component pictorially. Each view of each face is drawn out separately either full size or to the same scale. What is important is how we position the various views as this determines how we view and understand the drawing.

The views in the first angle projection shown in Figure 4.5 are arranged as follows:

- *Elevation.* This is the main view from which all the other views are positioned. You look directly at the side of the component and draw what you see.

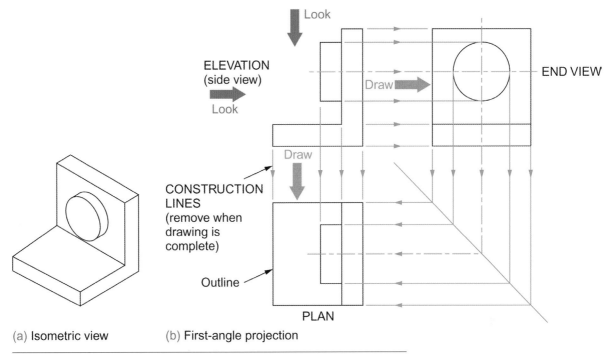

Figure 4.5 An isometric view and its corresponding first-angle projection.

(a) Isometric view   (b) First-angle projection

- *Plan.* To draw this, you look directly down on the top of the component and draw what you see below the elevation.
- *End view.* This is sometimes called an *end elevation*. To draw this you look directly at the end of the component and draw what you see at the opposite end of the elevation. There may be two end views, one at each end of the elevation, or there may be only one end view if this is all that is required to completely depict the component. Figure 4.5 requires only one end view. When there is only one end view this can be placed at either end of the elevation depending upon which gives the greater clarity and ease of interpretation. Whichever end is chosen, the rules for drawing this view must be obeyed.

Third-angle projection

Figure 4.6 shows the component in Figure 4.5a drawn in third angle projection. The views are arranged as follows:

- *Elevation.* Again, we have started with the elevation or side view of the component and, as you can see, there's no difference.
- *Plan.* Again, we look down on top of the component to see what the plan view looks like. However, this time, we draw the plan view above the elevation. This means that in third-angle projection we draw all the views from where we look.

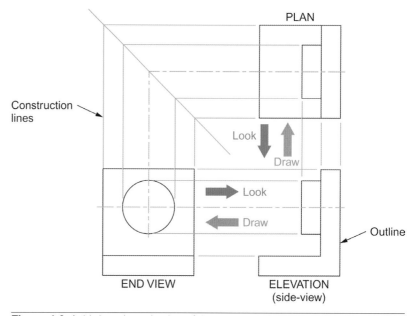

Figure 4.6 A third-angle projection of the isometric view shown in Figure 4.5a.

- *End view.* Note how the position of the end view is reversed compared with first-angle projection. This is because, like the plan view, we draw the end views at the same end from which we look at the component.

Auxiliary views

In addition to the main views on which we have just been working, we sometimes have to use auxiliary views. We use auxiliary views when we cannot show the true outline of the component or a feature of the component in one of the main views; for example, when a surface of the component is inclined as shown in Figure 4.7.

Test your knowledge 4.1

1 Figure 4.8 shows some parts drawn in first-angle projection and some in third-angle projection. Note that not all the possible views have been shown (only as many are shown as are actually needed). State which is first-angle and which is third-angle.

2 Two of the drawings in Figure 4.8 are symbols that are frequently included on drawing sheets in order to indicate whether a drawing is in first-angle or in third-angle. Which drawings do you think show these two symbols?

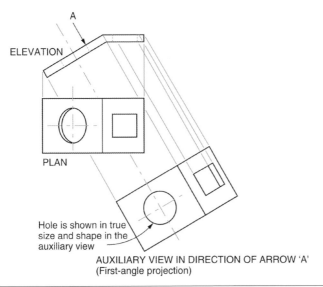

ELEVATION

PLAN

Hole is shown in true size and shape in the auxiliary view

AUXILIARY VIEW IN DIRECTION OF ARROW 'A'
(First-angle projection)

Figure 4.7 An auxiliary view used to clarify a drawing.

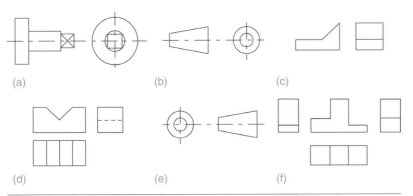

(a)

(b)

(c)

(d)

(e)

(f)

Figure 4.8 See Test your knowledge 4.1.

Activity 4.2

In Figure 4.9 each square has a side length of 10mm. Use a CAD package to draw the component in:

a) first-angle orthographic projection (only two views required)
b) third-angle orthographic projection (only two views required).

Activity 4.3

Obtain a small cast or machined component from your engineering workshop. Examine this carefully and then produce an orthographic drawing of the component showing separate elevation, plan and end elevation views. There is no need to add any detail such as dimensions and finishes at this stage.

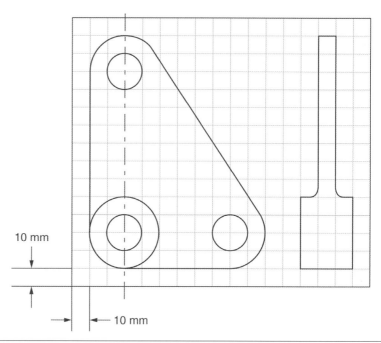

Figure 4.9 See Activity 4.2.

Activity 4.4

Use a CAD package to produce a first-angle projection of the component shown in Figure 4.10. Include dimensions in your drawing.

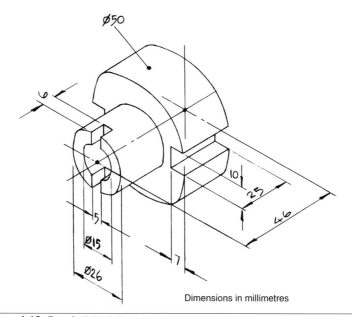

Dimensions in millimetres

Figure 4.10 See Activity 4.4.

Learning outcome 4.2

State the merits and limitations of different forms of communication: verbal, written, electronic means

The forms of communication that we use in everyday life can be broken down into four main types, namely:

- written
- graphical
- verbal
- other (non-verbal).

Each of these main type of communication can be further sub-divided. For example, graphical communication can take the form of drawings, sketches, block diagrams, exploded views, graphs, charts etc. Some of these can be further divided. For example, there are many different types of graph and chart. We have listed some of these in Figure 4.11.

In everyday life, we usually convey information by combining different forms of communication. For example, when we speak to other people we often combine verbal with non-verbal (body language) forms of communication. Body language can help add emphasis to our words or can be used to convey additional meaning. Presentations to groups of people usually involve verbal communication supported by visual aids such as PowerPoint slide presentations, handouts or flipcharts. Technical reports invariably combine written text with diagrams and photographs.

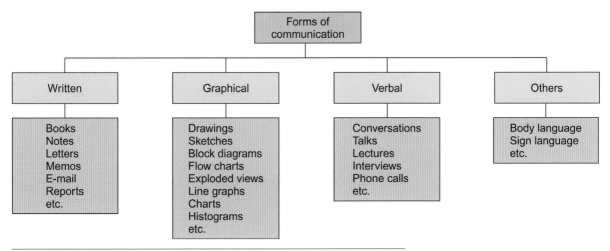

Figure 4.11 Forms of communication.

Electronic communication

Electronic communication, such as e-mail and file sharing, provides us with a means of quickly and efficiently conveying and exchanging information with other people. Conventional graphical images can be scanned and sent as document attachments or uploaded to file servers so that they can be accessed via a network or made available via the Internet. Electronic communication offers the following advantages:

- instant availability worldwide
- almost instantaneous transfer of information
- virtually unlimited space available for storage (documents and images can be very large)
- ability to work collaboratively with other people
- information can be encrypted for security
- information can be updated easily.

Test your knowledge 4.2

State **four** advantages of using electronic means of communication.

Test your knowledge 4.3

From the types of communication shown in Figure 4.11, identify (with reasons) methods of communication that you consider most appropriate in each of the following situations:

a) making an appointment to see a doctor
b) apologizing for forgetting your sister's birthday
c) directing a friend to a restaurant in the next town
d) selling your car

Present your work in the form of a brief set of handwritten notes.

Test your knowledge 4.4

Complete Table 4.1 by placing a tick in the box against the applicability of different forms of communication in relation to fitting a plug to the electric cable on a portable appliance.

Table 4.1 See Test your knowledge 4.4.

Situation	Highly applicable	Possibly applicable	Not applicable
A written instruction sheet			
A verbal commentary supplied in mp3 format			
A sequence of diagrams with brief text			
A chart that lists each of the steps required			
A YouTube video			

Information sources

Engineers use a wide variety of information in their everyday lives. This information is derived from a variety of different sources including:

- books
- application notes
- technical reports
- data sheets and data books
- catalogues
- engineering drawings
- CD-ROMs
- databases
- websites.

Books

Books, whether they are paper-based publications or one of the new generation of electronic books (*e-books*) provide information on an almost infinite number of subjects and a good technical library can be invaluable in any engineering context. All books contain summary information. This typically including the date of first publication, information concerning copyright, and an ISBN number.

When using a book as a source of information it is important to ensure that it is up to date. It is also necessary to ensure that the content is reliable and that there are no omissions or errors. Book reviews (often published online and in the technical press) can be useful here!

Test your knowledge 4.5

Take a look at the information that appears in the first few pages of this book and then answer the following questions:

1 What was the date of first publication?
2 Is this book a reprint?
3 Who owns the copyright?
4 What is the book's ISBN number?
5 Who has published the book?

Application notes

> **Key point**
>
> Application notes explain how something is used in a particular application or how it can be used to solve a particular problem. Application notes are intended as a guide for designers and others who may be considering using a particular process or technology for the first time. Technical reports, on the other hand, provide information that is more to do with whether a component or device meets a particular specification or how it compares with other solutions. Technical reports are thus more useful when it comes to analysing how a process or technology performs than how it is applied.

Application notes are usually brief notes (often equivalent in length to a chapter of a book) supplied by manufacturers in order to assist engineers and designers by providing typical examples of the use of engineering components and devices. An application note can be very useful in providing practical information that can help designers to avoid pitfalls that might occur when using a component or device for the first time.

Technical reports

Technical reports are somewhat similar to application notes but they focus more on the performance specification of engineering components and devices (and the tests that have been carried out on them) than the practical aspects of their use. Technical reports usually include detailed specifications, graphs, charts and tabulated data. Technical reports are produced by manufacturers as well as industry groups and end users of engineering products.

Data sheets and data books

> **Key point**
>
> Data sheets provide concise information on parts, components and materials. Data sheets provide sufficient information to allow you to identify a part and know what it is designed to do.

Data sheets usually consist of abridged information on a particular engineering component or device. They usually provide maximum and minimum ratings, typical specifications, as well as information on dimensions, packaging and finish. Data sheets are usually supplied free on request from manufacturers and suppliers. Collections of data sheets for similar types of engineering components and devices are often supplied in book form. Often supplementary information is included relating to a complete family of products. An example of a data sheet is shown in Figure 4.12.

Test your knowledge 4.6

Refer to the extract from the data sheet shown in Figure 4.12 and use it to answer the following questions:

Howard Associates

BJ284
NPN Silicon Power Transistor

MAIN FEATURES
- 150W max. power dissipation
- High current gain (>100 typ. at I_C = 2A)
- Large gain-bandwidth product (>6 MHz typ. at I_C = 2A)
- Rugged TO3 case
- Complementary to BJ285 PNP transistor

APPLICATIONS
- High-quality audio power amplifiers
- Linear voltage regulators
- Power switching
- Automotive ignition systems
- Power control and regulation
- Emergency lighting systems

ABSOLUTE MAXIMUM RATINGS (T_A = +25°C)

Collector to base voltage, V_{CBO}	180V
Collector to emitter voltage, V_{CEO}	180V
Emitter to base voltage, V_{EBO}	5V
Collector current, I_C	16A
Emitter current, I_C	16A
Power dissipation, P_C	150W
Junction temperature, T_j	+150°C
Storage temperature, T_{stg}	-65°C to +150°C

SYMBOL

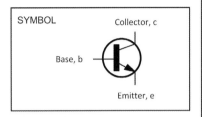

Collector, c
Base, b
Emitter, e

TO3 PACKAGE

ELECTRICAL CHARACTERISTICS (T_A = +25°C)

Parameter	Symbol	Condition	Min.	Typ.	Max.	Unit
Collector cut-off current	I_{CBO}	V_{CB} = 90V, I_E = 0	–	–	100	µA
Emitter cut-off current	I_{EBO}	V_{EB} = 5V, I_C = 0	–	–	100	µA
DC current gain	h_{FE}	V_{CE} = 5V, I_C = 2A	70	–	140	–
Collector-emitter saturation voltage	$V_{CE(sat)}$	I_C = 10A, I_B = 1A	–	–	3.0	V
Base to emitter voltage	V_{BE}	V_C = 5V, I_C = 10A	–	–	2.5	V
Current gain bandwidth product	f_T	V_C = 5V, I_C = 2A	–	6	–	MHz
Output capacitance	C_{ob}	V_{CB} = 10V, f = 1MHz	–	300	–	pF

CASE DIMENSIONS

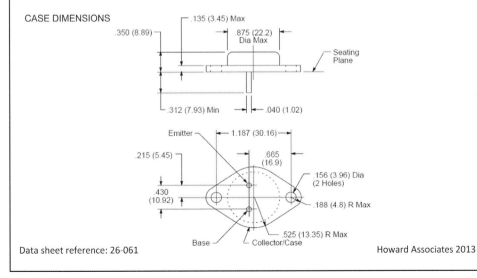

Data sheet reference: 26-061

Howard Associates 2013

Figure 4.12 Example of a data sheet.

1 Who is the manufacturer of the BJ284 device?
2 How many pin connections does the device have?
3 What type of transistor is a BJ284?
4 What is the maximum power dissipation for a BJ284 transistor?
5 State three applications of a BJ284 transistor.
6 What is the maximum junction temperature for a BJ284 transistor?
7 Under what conditions is the DC current gain specified?
8 A BJ284 is to be operated with a collector-emitter voltage of 6V and a collector current of 20A. Is this within the ratings for the device? Give reasons for your answer.

Catalogues

Most manufacturers and suppliers provide catalogues that list their full product range. These often include part numbers, illustrations, brief specifications and prices. While catalogues are often extensive documents with many hundreds or thousands of pages, short-form catalogues are usually also available. These usually just list part numbers, brief descriptions and prices but rarely include any illustrations. A brief extract from a short-form catalogue is shown in Figure 4.13.

Diecast Boxes   IP65 Sealed/Painted

A range of high-quality diecast aluminium boxes with an optional grey epoxy paint finish to RAL7001. The lid features an integral synthetic rubber sealing gasket and captive stainless steel fixing screws. Mounting holes and lid fixing screws are outside the seal, giving the enclosure protection to IP 65.

Standard supply multiple = 1   Delivery normally ex-stock

| Size | | | | | | | Price each | | |
L	W	H	T	Finish	Manufacturer's ref:	Stock code	1-9	10-24	25+
90	45	30	3.0	none	1770-1541-21	DB65-01	£4.52	£3.95	£3.50
90	45	30	3.0	grey	1770-1542-21	DB65-01P	£5.40	£4.90	£4.45
110	50	30	4.5	none	1770-1543-22	DB65-02	£5.25	£4.50	£4.15
110	50	30	4.5	grey	1770-1544-22	DB65-02P	£6.42	£5.37	£4.95
125	85	35	5.0	none	1770-1545-23	DB65-03	£6.15	£5.17	£4.71
125	85	35	5.0	grey	1770-1546-23	DB65-03P	£7.10	£6.05	£5.65

Figure 4.13 Example of an extract from a short-form catalogue.

Test your knowledge 4.7

Use the catalogue extract shown in Figure 4.13 to answer the following questions:

1 What type of fixing screws are supplied?
2 What are the dimensions of the largest box in the series?
3 Is this type of box suitable for use where there is a high degree of humidity? Explain your answer.
4 Determine the total cost of a batch of 15 grey paint finished boxes each measuring 90 × 45 × 30mm.

Activity 4.5

Your company requires 15 diecast boxes suitable for enclosing a printed circuit board of thickness 3mm measuring 80mm × 35mm. The tallest component stands 15mm above the board and a minimum clearance of 5mm is to be allowed around the board. The enclosure is to be supplied ready for mounting the printed circuit board and should not need any further finishing other than drilling. Prepare an e-mail message to the Sales Department at Dragon Components (the hardware supplier whose short-form catalogue extract appears in Figure 4.12) and make sure that it gives all the information required to fulfil your order.

Activity 4.6

Prepare and deliver a brief verbal presentation of no more than ten minutes on any one of the following topics:

1 modern racing bikes
2 the Raspberry Pi computer
3 how to change a car tyre
4 how to fly a low-cost drone
5 the latest British Navy aircraft carrier

Prepare a single A4 page written handout for your audience and include full details of all the information sources used.

Learning outcome 4.3

Identify basic drawing conventions, layouts and the use of sketches

A number of conventions need to be followed when preparing an engineering drawing. In order to help other people understand exactly what a drawing represents a set of precise rules needs to be followed. For example, when a line is drawn we need to know whether this represents something that is immediately visible rather than being hidden from view. Engineering drawing conventions are widely understood because they are governed by standards defined by national bodies such as the British Standard Institution (BS 8888:2013) and the American National Standards Institute (Y14 series).

Planning an engineering drawing

Before you begin to draw anything you need to have a plan. This can help you avoid having to make changes as you go along and, in an extreme case, having to start from scratch again. You first need to decide what type of drawing is required and whether it is to be a sketch, a pictorial diagram, an orthographic drawing like the ones that you met in Section 4.1, or whether it should be some form of schematic diagram. If orthographic, you need to decide on the projection that you are going to use and whether you will need any auxiliary drawings to clarify any particular features of the drawing.

Paper size

When you start to plan a drawing that will be printed you will need to decide on the paper size that will eventually be used. Conventional printed drawings are usually produced on 'A' size paper. Paper size A0 is approximately one square metre in area and is the basis of the system. Size A1 is half the area of size A0, size A2 is half the area of size A1 and so on down to size A4. Smaller sizes are available but they are not used for drawing. All the 'A' size sheets have their sides in the ratio of 1:1.2. This gives the following paper sizes (see Figure 4.14):

- A0 = 841mm × 1189mm
- A1 = 594mm × 841mm
- A2 = 420mm × 594mm
- A3 = 297mm × 420mm
- A4 = 210mm × 297mm.

The paper size you choose will depend upon the size of the drawing and the number of views required. Be generous – nothing looks worse than a cramped up drawing and overcrowded dimensions. It is also a false economy since overcrowding invariably leads to reading errors.

As you will already have seen from some of the previous examples, the drawing should always have a border and a title block. This restricts the blank area available to draw on. Figures 4.15 and 4.16

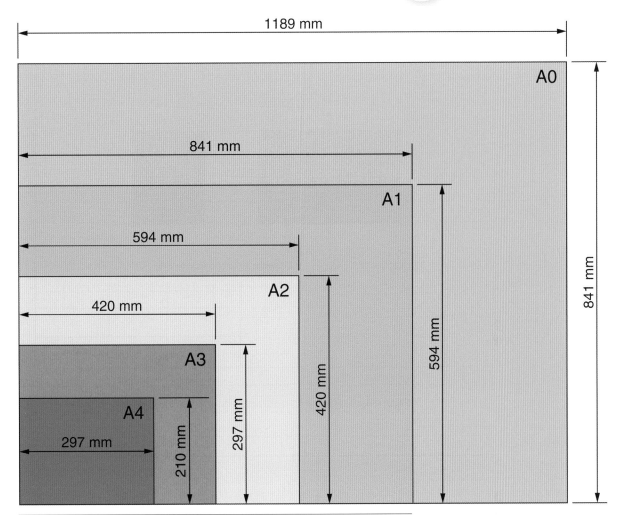

Figure 4.14 Standard paper sizes.

show how the views should be positioned. These layouts are only a guide but they offer a good starting point until you become more experienced. If only one view is required then it is centred in the drawing space available.

Title block

Hopefully, you've noticed that the drawings shown in Figures 4.15 and 4.16 include a *title block* at the bottom of the page. When studying an engineering drawing the title block is the first thing that you should read. The title block should usually contain:

- the drawing number (which should be repeated in the top left-hand corner of the drawing)
- the drawing name (title)
- the drawing scale

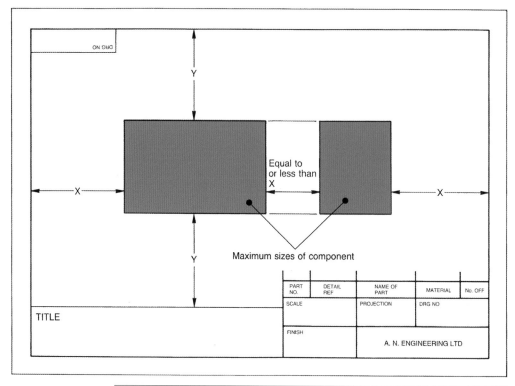

Figure 4.15 Positioning a drawing with two component parts.

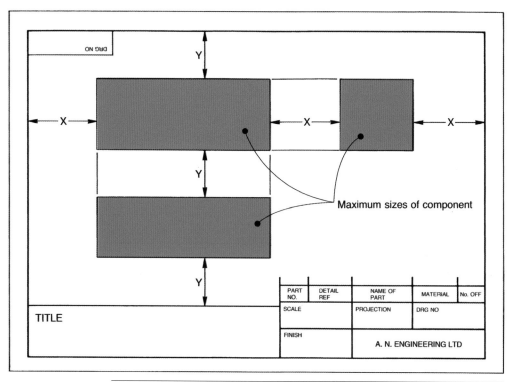

Figure 4.16 Positioning a drawing with three component parts.

- the projection used (standard symbol)
- the name and signature of the person who made the drawing together with the date on which the drawing was made
- the name and signature of the person who checked and/or approved the drawing, together with the date of signing
- the issue number and its release date
- any other information as dictated by company policy.

Scale

It is unusual for a drawing to be made that's life-size. Instead, the dimensions are scaled so that the drawing will fit comfortably on the page (as shown previously in Figures 4.15 and 4.16). The scale should always be stated on the drawing as a ratio. The recommended scales are as follows:

- Full size = 1:1

- Reduced scales (smaller than full size) are:

  1:2 1:5 1:10
  1:20 1:50 1:100
  1:200 1:500 1:1000
  (NEVER use the words full-size, half-size, quarter-size etc.).

- Enlarged scales (larger than full size) are:

  2:1 5:1 10:1
  20:1 50:1 100:1

Screw threads

Standard conventions are used in order to avoid having to draw out, in detail, common features in frequent use. For example, Figure 4.17a shows a pictorial representation of a screw thread. Figure 4.17b shows the convention for a screw thread. The convention for the screw thread is much the quicker and easier to draw. In the next section we will be looking at the conventions used for drawing lines, showing details and including dimensions in more detail.

Key point

Standard drawing conventions help to ensure that engineering drawings are unambiguous and widely understandable.

Key point

Before you begin to draw anything you need to have a plan. You first need to decide what type of drawing is required and whether it should be a sketch, a pictorial diagram, an orthographic drawing or a schematic diagram. If it is an orthographic drawing you need to decide which projection (either first-angle or third-angle) is to be used.

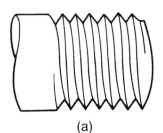
(a)

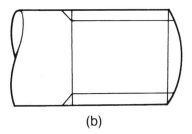
(b)

Figure 4.17 Drawing convention for showing a screw thread.

Test your knowledge 4.8

Explain why drawing conventions are important. What national body is responsible for drawing standards and conventions in the UK?

Test your knowledge 4.9

List **four** features that you would expect to see in the title block of a drawing.

Test your knowledge 4.10

Why is the scale of an engineering drawing important and how is it shown on a formal engineering drawing?

Learning outcome 4.4

Identify types of lines, detailing and dimensioning

Lines

A number of drawing conventions apply specifically to the way in which lines are shown on a formal engineering drawing. In all cases the lines of a drawing should be uniformly black, dense and bold. Lines should be thick or thin as recommended later. Thick lines should usually be twice as thick as thin lines. Figure 4.18 shows the types of lines recommended in BS 8888:2013 for use in engineering drawing and how the lines should be used (see Figure 4.19).

Sometimes the lines overlap in different views. When this happens the following order of priority should be observed:

1 Visible outlines and edges (type A) take priority over all other lines.
2 Next in importance are hidden outlines and edges (type E).
3 Then cutting planes (type G).
4 Next come centre lines (type F and B).
5 Outlines and edges of adjacent parts etc. (type H).
6 Finally, projection lines and shading lines (type B).

Line type	Example	Description	Application
A	——————	Continuous thick	Visible outlines and edges
B	———————	Continuous thin	Dimension, projection and leader lines, hatching, outlines of revolved sections, short centre lines, imaginary intersections
C	∿∿∿	Continuous thin irregular	Limits of partial or interrupted views and sections
D	—ᐱ—ᐱ—ᐱ—	Continuous thin with zigzags	Limits of partial or interrupted views and sections
E	— — — — — —	Dashed thin	Hidden outlines and edges
F	— · — · — · —	Chain thin	Centre lines, lines of symmetry, trajectories and loci, pitch lines and pitch circles
G	⌐· — · — ·⌐	Chain thin, thick at ends and changes of direction	Cutting planes
H	— ·· — ·· — ·· —	Chain thin double dashed	Outlines and edges of adjacent parts, outlines and edges of alternative and extreme positions of movable parts, initial outlines prior to forming, bend lines on developed blanks or patterns

Figure 4.18 Types of line.

Figures 4.19 and 4.20 show some examples of how the rules concerning lines should be applied:

- In Figure 4.19a hatching (type B) is used inside a continuous area (type A).
- In Figure 4.19b the limit of view is indicated (type C).
- In Figure 4.19c the centre lines have been marked (type F).
- In Figure 4.19d the hidden detail is shown (type E) when the part is viewed from the side (note that the left-hand diagram is a plan view while the right-hand diagram is the corresponding side view).
- In Figure 4.20 the component shown is able to move from its resting position A to position B. Its extreme position is shown using line type H.

Key point

Visible outlines take the highest priority when placing lines on an engineering drawing.

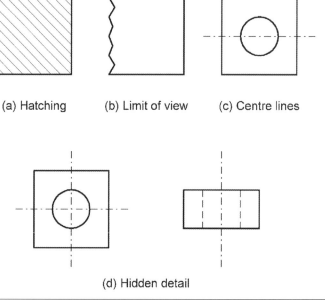

(a) Hatching   (b) Limit of view   (c) Centre lines

(d) Hidden detail

Figure 4.19 Use of various line types.

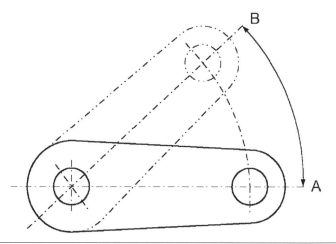

Figure 4.20 Indicating extreme positions.

Leader lines

Leader lines, as their name implies, lead written information or dimensions to the points where they apply. Leader lines are thin lines (type B) and they end in an arrowhead or in a dot, as shown in Figure 4.21a. Arrowheads touch and stop on a line, while dots should always be used within an outline.

• When an arrowed leader line is applied to an arc it should be in line with the centre of the arc, as shown in Figure 4.21b.

- When an arrowed leader line is applied to a flat surface, it should be nearly normal to the lines representing that surface, as shown in Figure 4.21c.
- Long and intersecting leader lines should not be used, even if this means repeating dimensions and/or notes, as shown in Figure 4.21d.
- Leader lines must not pass through the points where other lines intersect.
- Arrowheads should be triangular with their length some three times larger than the maximum width. They should be formed from straight lines and the arrowheads should be filled in. The arrowhead should be symmetrical about the leader line, dimension line or stem. It is recommended that arrowheads on dimension and leader lines should be some 3 to 5mm long.
- Arrowheads showing direction of movement or direction of viewing should be around 7 to 10mm long. The stem should be the same length as the arrowhead or slightly greater. It must never be shorter.

Key point

Leader lines should not be too long and should not intersect with other lines.

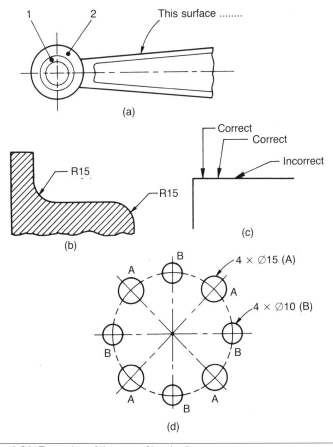

Figure 4.21 Examples of the use of leader lines.

Test your knowledge 4.11

Identify each of the line styles shown in Figure 4.22.

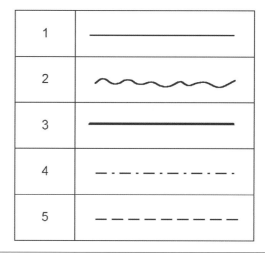

Figure 4.22 See Test your knowledge 4.11.

Test your knowledge 4.12

Figure 4.23 shows a simple drawing using a variety of different lines. State whether each numbered line is correctly or incorrectly marked and, if incorrect, how the line should be corrected.

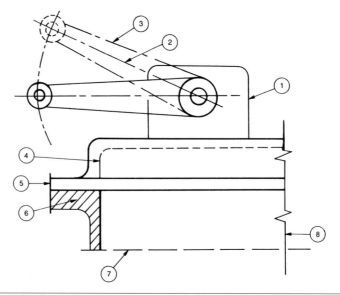

Figure 4.23 See Test your knowledge 4.12.

Test your knowledge 4.13

Figure 4.24 shows some leader lines marked with arrowheads and others marked with dots. State whether each numbered line is correctly or incorrectly marked and, if incorrect, how the line should be corrected.

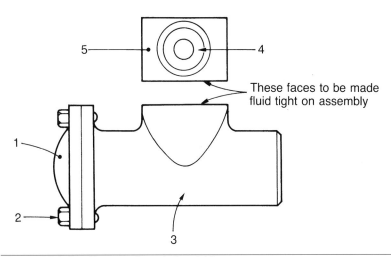

These faces to be made fluid tight on assembly

Figure 4.24 See Test your knowledge 4.13.

Letters and numerals

Style

The style should be clear and free from embellishments. In general, capital letters should be used. A suitable style could be:

ABCDEFGHIJKLMNOPQRSTUVWXYZ

1234567890

Size

The characters used for dimensions and notes on drawings should be not less than 3mm tall. Title and drawing numbers should be at least twice as big.

Direction of lettering

Notes and captions should be positioned so that they can be read in the same direction as the information in the title block. Dimensions have special rules and will be dealt with later.

Location of notes

General notes should all be grouped together and not scattered about the drawing. Notes relating to a specific feature should be placed adjacent to that feature.

Emphasis

Characters, words and/or notes should not be emphasized by underlining. Where emphasis is required the characters should be enlarged.

Detailing

Detailing a drawing by adding text (dimensions, notes, datums, geometric tolerances, and other annotation) can help make it more useful and less liable to misinterpretation. Detailing is often added only after the basic drawing has been produced. The aim of detailing is not only to clarify aspects of the drawing, but also to provide additional information that might not be apparent from the basic drawing. Typical examples might be to include the required surface finish or the tolerance required of a finished manufactured part.

Symbols and abbreviations

If all the detail required for a drawing had to be written out in full, the drawing would quickly become very cluttered and so symbols and abbreviations are often used to simplify text and notes. Wherever possible, abbreviations should be avoided but if they do have to be used they need to conform with relevant standards such as BS ISO 80000. Note also that abbreviations (text equivalents) shall be the same in the singular and plural and full stops should not be used except where the abbreviation forms a word (e.g. No. as an abbreviation for 'number').

> **Key point**
>
> Detailing is used to clarify an engineering drawing by adding text and other useful information such as dimensions, tolerances and finishes.

Activity 4.7

Refer to relevant standards and other documentation and use it to complete Table 4.2.

Activity 4.8

Refer to relevant standards and other documentation and use it to complete Figure 4.25.

Table 4.2 See Activity 4.7.

Abbreviation or symbol	Meaning
ASSY	
BS	
CL	
	Countersink
Ø	
	Diameter (in a note)
LH	
	Maximum
	Minimum
NTS	
	Number
R	
RH	
SPEC	
	Tolerance
VOL	

Dimensioning

When a component is being dimensioned, the dimension lines and the projection lines should be thin full lines (type B). Where possible, dimensions should be placed outside the outline of the object, as shown in Figure 4.26a. The rules are:

- Outline of object to be dimensioned in thick lines (type A).
- Dimension and projection lines should be half the thickness of the outline (type B). There should be a small gap between the projection line and the outline.
- The projection line should extend to just beyond the dimension line.
- Dimension lines end in an arrowhead that should touch the projection line to which it refers.
- All dimensions should be placed in such a way that they can be read from the bottom right-hand corner of the drawing.

The purpose of these rules is to allow the outline of the object to stand out prominently from all the other lines and to prevent confusion.

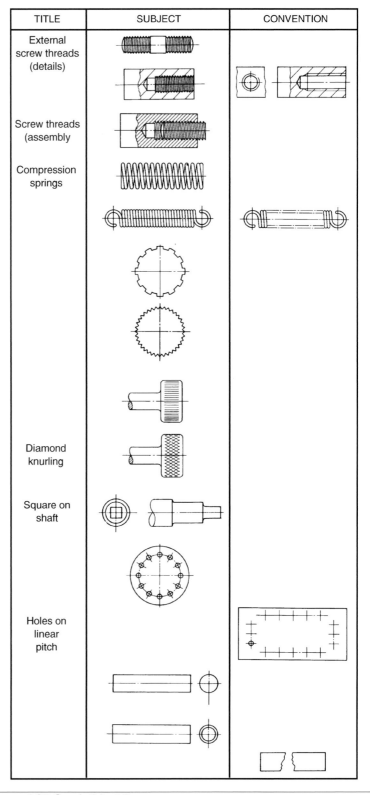

TITLE	SUBJECT	CONVENTION
External screw threads (details)		
Screw threads (assembly		
Compression springs		
Diamond knurling		
Square on shaft		
Holes on linear pitch		

Figure 4.25 See Activity 4.8.

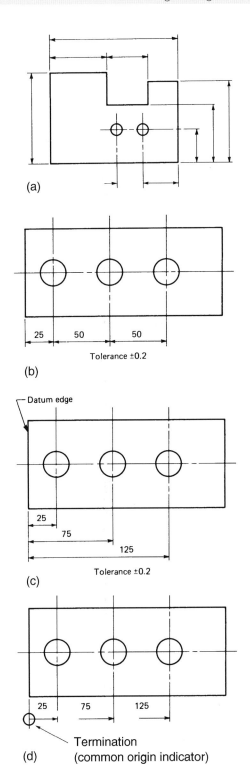

(a)

(b)

Tolerance ±0.2

25   50   50

(c)

Datum edge

25
75
125

Tolerance ±0.2

(d)

25   75   125

Termination
(common origin indicator)

Figure 4.26 Dimensioning.

There are three ways in which a component can be dimensioned. These are:

- Chain dimensioning, as shown in Figure 4.26b.
- Absolute dimensioning (dimensioning from a datum) using parallel dimension lines, as shown in Figure 4.26c.
- Absolute dimensioning (dimensioning from a datum using superimposed running dimensions, as shown in Figure 4.26d). Note the common origin (termination) symbol.

It is neither possible to manufacture an object to an exact size nor to measure an exact size. Therefore important dimensions have to be *toleranced*. That is, the dimension is given two sizes: an upper limit of size and a lower limit of size. Providing the component is made so that it lies between these limits it will function correctly. Information on limits and fits can be found in BS 4500:1969.

The method of dimensioning can also affect the accuracy of a component and produce some unexpected effects. Figure 4.26b shows the effect of chain dimensioning on a series of holes or other features.

The designer specifies a common tolerance of ±0.2 mm. However, since this tolerance is applied to each and every dimension, the cumulative tolerance becomes ±0.6 mm by the time you reach the final, right-hand hole, which is not what was intended. Therefore, absolute dimensioning as shown in Figure 4.26c and 4.26d is to be preferred in this example.

With absolute dimensioning, the position of each hole lies within a tolerance of ±0.2 mm and there is no cumulative error. Further examples of dimensioning techniques are shown in Figure 4.27.

Activity 4.9

Figure 4.28 shows a component drawn in isometric projection. Redraw the component using first-angle orthographic projection and include dimensions.

Sectioning

Sectioning is used to show the hidden detail inside hollow objects more clearly than can be achieved using dashed thin (type E) lines. Figure 4.29a shows an example of a simple sectioned drawing. The cutting plane is the line A–A. In your imagination you remove everything to the left of the cutting plane, so that you

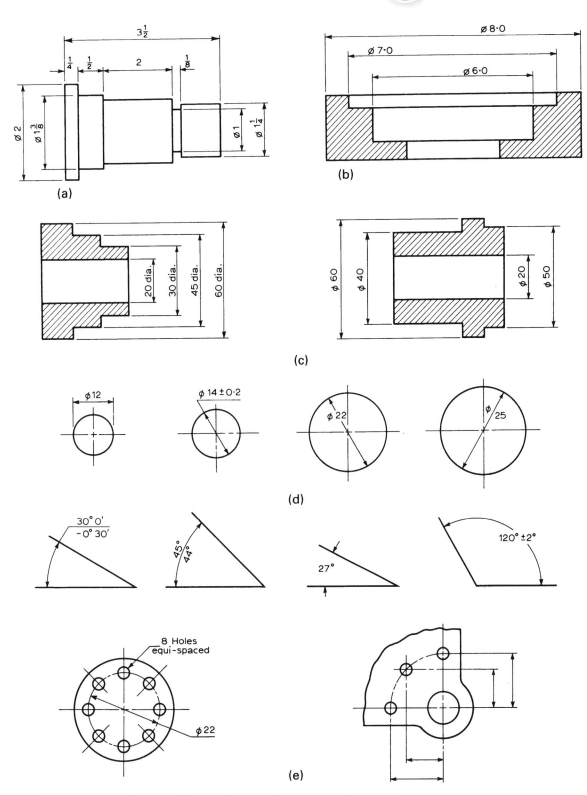

Figure 4.27 Some more examples of dimensioning.

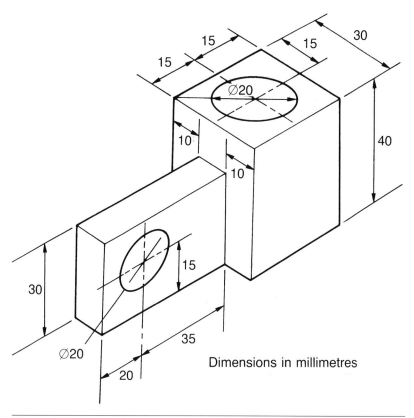

Figure 4.28 See Activity 4.9.

only see what remains to the right of the cutting plane, looking in the direction of the arrowheads. Another example is shown in Figure 4.29b.

When interpreting sectioned drawings, some care is required. It is easy to confuse the terms *sectional view* and *section*. Figure 4.29c shows how to section an assembly. Note how solid shafts and the key are not sectioned. Also note that thin webs that lie on the section plane are not sectioned.

Sectional views

In a sectional view you see the outline of the object at the cutting plane. You also see all the visible outlines seen beyond the cutting plane in the direction of viewing. Therefore, Figure 4.29a is a sectional view.

Sections

A section only shows the outline of the object at the cutting plane. Visible outlines beyond the cutting plane in the direction of viewing are not shown. Therefore, a section has no thickness.

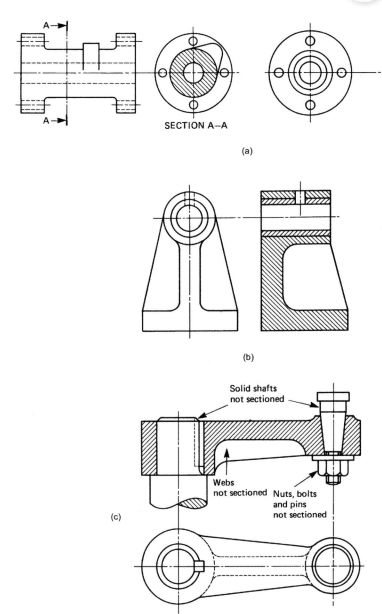

SECTION A–A

(a)

(b)

Solid shafts
not sectioned

Webs
not sectioned   Nuts, bolts
and pins
not sectioned

(c)

Figure 4.29 Sectioning and hatching.

Cutting planes

You have already been introduced to cutting planes in the previous
examples. They consist of type G lines; that is, a thin chain line
that is thick at the ends and at changes of direction. The direction
of viewing is shown by arrows with large heads. The points of the
arrowheads touch the thick portion of the cutting plane. The cutting
plane is labelled by placing a capital letter close to the stems of
the arrows. The same letters are used to identify the corresponding
section or sectional view.

Hatching

You will have noticed that the shading of sections and sectional views consists of sloping, thin (type B) lines. This is called hatching. The lines are equally spaced, slope at 45° and are not usually less than 4mm apart. However, when hatching very small areas the hatching can be reduced, but never less than 1mm. The drawings in this book may look as though they do not obey these rules. Remember that they have been reduced from much bigger drawings to fit onto the pages.

Test your knowledge 4.14

Explain the purpose of sectioning parts of a drawing.

Test your knowledge 4.15

How is a cutting plane identified in a sectioned drawing?

Review questions

1. Give **four** advantages of using CAD to create an engineering drawing when compared with manual drawing techniques.

2. Explain what is meant by orthographic drawing. How does this differ from isometric drawing?

3. What is an auxiliary view and in what circumstances would you use one?

4. Give **one** example of **each** of the following types of communication:
   a) graphical
   b) written
   c) verbal
   d) electronic

5. List **six** different information sources that engineers use regularly.

6. Briefly explain the difference between an application note and a technical report.

7. What information would you expect to find in a data sheet for a rechargeable battery? List at least **five** important features.

8. An engineering drawing has three separate components shown on the same drawing sheet. Use a sketch to show how they should be positioned on the sheet.

9. Explain, with the aid of a sketch, the drawing convention used for indicating a screw thread.

10. When placing lines on an engineering drawing, which type of line has the highest order of priority and how is it shown on the drawing?

11. Explain why tolerancing is used when adding dimensions to a drawing and give an example of how this is shown on a typical detailed drawing.

12. Explain, with the aid of a sketch, the difference between chain and absolute dimensioning.

Chapter checklist

Learning outcome	Page number
4.1 Draw an orthographic drawing.	58
4.2 State the merits and limitations of different forms of communication: • verbal • written • electronic means.	67
4.3 Identify basic drawing conventions, layouts and the use of sketches.	73
4.4 Identify types of lines, detailing and dimensioning.	78

Engineering applications of IT and ICT

Chapter summary

Information and Communications Technology (ICT) has had a huge impact on the engineering industry. Not only have these exciting new technologies revolutionized the way that we design and manufacture products but they have also given us a way of communicating and sharing information quickly and easily. The prime mover in all of this has been the availability of increasingly small and powerful computer systems coupled with the widespread use of digital communication via wired and wireless networks and the Internet. These new technologies have been made possible by advances in microelectronics coupled with advances in software engineering. We shall begin by introducing you to some of the more important items of IT/ICT hardware before moving on to describe some applications of IT in the engineering sector.

Learning outcome 5.1

Identify the types of IT and ICT hardware used in engineering applications

Many engineering tasks are assisted by systems that use IT and ICT to solve problems and perform tasks that would be difficult or even impossible to solve using manual techniques. IT and ICT involves hardware and software working together, often exchanging data and information over some considerable

Figure 5.1 Engineers rely on IT and ICT to perform everyday tasks.

distance. For example, it is possible to use satellites and the Internet to continuously track the position of an aircraft anywhere in the world. If an emergency occurs or if the aircraft develops an engineering problem the owner or operator can take immediate action.

Hardware is a generic term that's applied to the physical parts of a computer, in other words parts that you can touch, hold and move around. Hardware is often classified as being internal or external. *Internal hardware* applies to items that are found inside the physical enclosure in which a computer is placed. *External hardware* exists outside a computer's physical enclosure but must be connected to it in some way. Because it lies outside the boundary or periphery of a computer system, external hardware is often also referred to as *peripheral hardware*. In many cases wires and cables are used to link internal and external hardware, but increasingly wireless interconnection is being used. *Wireless* connection involves the use of radio waves to convey digital signals between computers and peripherals.

Key point

Hardware can be either internal or external depending on whether it is located inside or outside a computer system's enclosure.

Internal hardware

Examples of internal hardware include:

- a central processing unit (the 'brains' of the computer)
- memory that can be used for storing data on a temporary basis (such as semiconductor read/write memory)
- memory that can be used for permanently storing data (such as semiconductor read-only memory)
- one or more disk or optical *drives* for holding large amounts of data
- a clock that provides a time reference for the system
- a graphics controller for processing output for a display device
- other controllers, as required.

The internal components of a typical computer are shown in Figure 5.2. The central processing unit (CPU) is responsible for managing the system as well as fetching, decoding and executing program instructions. The CPU also has an internal arithmetic logic unit that is able to perform mathematical operations such as addition and subtraction as well as logical operations such as AND and OR. The CPU requires a clock in order to provide a master timing reference for the system which ensures that all movements of data around the system are synchronized.

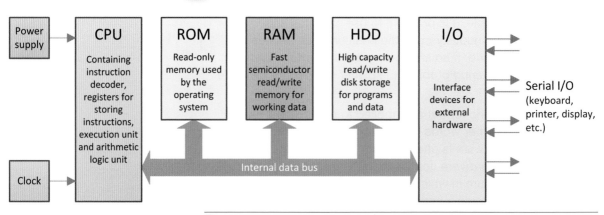

Figure 5.2 Internal components of a typical computer system.

Test your knowledge 5.1

Name **three** tasks performed by a computer's central processing unit (CPU).

External hardware

Examples of external hardware (see Figure 5.3) include:
- screens and other displays
- keyboards for inputting text, numbers and other *characters*

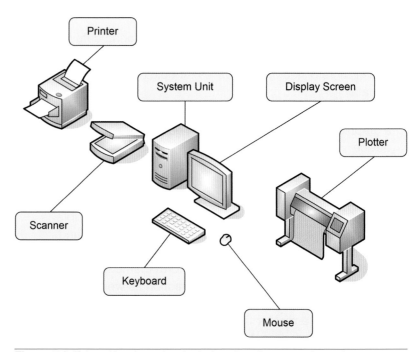

Figure 5.3 External hardware in a typical engineering computer system.

- pointing devices such as mice or trackballs
- printers to provide hard copy output
- external drives for storing large amounts of data
- routers to provide network access.

Test your knowledge 5.2

The devices that we've listed under the general heading of 'external hardware' fall into three categories according to whether they act as inputs, outputs or both. Classify each item in the list of 'external hardware' under one of these three headings.

Microprocessors

Microprocessors are essentially computers where the functions of the central processing unit (CPU) are combined within a single integrated circuit or 'silicon chip'. The first microprocessors were developed in the early 1970s. These were simple and crude by today's standards but they found an immediate application in the automotive industry where they were deployed in engine management and automatic braking systems, and later in the first generation of personal computers which manipulated data 8-bits at a time. Modern computers are capable of operating on 32 or 64 bits at a time at speeds that are more than 100 times faster than first-generation systems. Coupled with a vastly increased memory capacity and the ability to store large amounts of data using hard disk drives (HDD) and CD/DVD drives, modern computers have become invaluable in an engineering environment.

Key point

Microcomputers can manipulate large amounts of data very quickly and this makes them suitable for performing complex engineering tasks such as computer-aided engineering (CAE), computer-aided manufacture (CAM) and computer numerical control (CNC).

Microcontrollers

Microcontrollers, like the one shown in Figure 5.4, are dedicated microcomputer systems that perform specific tasks such as operating a robot arm or controlling an air-conditioning system. The tasks performed by most microcontrollers are not particularly demanding nor are they time-critical and, because of this, they don't require large amounts of memory nor do they need to operate at the very high speeds needed with conventional desktop computers. Microcontrollers have input and output (I/O) connections and *bus systems* that allow them to be interfaced directly with motors, actuators, sensors and a wide variety of other components needed for controlling external hardware.

Key point

Microcontrollers are small computers dedicated to specific tasks. They often have limited memory capability but provide input/output (I/O) that makes them suitable for connecting directly to a wide range of external hardware, such as keypads, motors and actuators.

Figure 5.4 An Arduino microcontroller. The large chip contains the CPU and memory. External devices can be connected using the digital and analogue I/O connectors located around the edge of the printed circuit board (PCB). The board is programmed using a universal serial bus (USB) connector.

Key point

Microcomputers and microcontrollers communicate with external devices by using one or more standard bus systems. These systems operate according to established communication protocols and they help to simplify the wiring and interconnection of I/O devices.

Test your knowledge 5.3

What do the following abbreviations stand for?

a)   CAD

b)   CAM

c)   CNC

d)   USB

e)   PCB

f)   I/O

Test your knowledge 5.4

State **three** typical applications of microcontrollers.

Test your knowledge 5.5

State **three** differences between a microcontroller and a general-purpose desktop microcomputer system.

Activity 5.1

Locate a computer system that's used for an engineering application (such as CAD or CAM) in your company or training centre. Investigate the system and identify each of the external hardware components of the system and list them together with a brief description of each of them. Create a block diagram of the system showing the links between components. Label your drawing clearly.

Learning outcome 5.2

Identify the types of IT and ICT operating systems used in engineering applications

A computer needs software as well as hardware in order to do anything. Software provides you with a means of controlling the hardware by sending instructions and data to it. For example, to be able to input text from a keyboard the computer needs to be programmed so that it responds to a key press, identifies the key that's being pressed and then acts on it. Unlike hardware, software cannot be 'seen'. Instead, it exists as a series of instructions and data held in digital form as a sequence of 1s and 0s in the computer's memory.

There are several different types of software which you need to know about, including:

- operating systems
- utility programs
- application programs.

Operating systems (sometimes referred to as *system software*) perform several important functions, such as providing:

- a low-level interface to the computer system's hardware
- a file system that manages the storage of programs and data
- an interface that will handle user input and output.

To make life easier for users, most modern operating systems incorporate a graphical user interface (GUI). This usually supports a means of sending commands to the computer by controlling a pointer or touching a screen. Examples of the most popular modern operating systems include Windows, Linux, Unix, Mac OS X and Android.

Key point

In order to provide us with
a fully functioning computer
system we need operating
system software capable
of managing the various
hardware resources as well as
providing us with an interface
so that we can interact with
the system and make use of
applications software such as
text editors, spreadsheets,
database managers and CAD
packages.

Figure 5.5 shows how the operating system manages the interface between one or more user application programs and the hardware components of the system, together with any peripheral devices (such as keyboards, displays and printers) that are connected to it. The operating system operates at both high level, providing a variety of *system services*, such as those needed to operate a *file system*, and at low level, using a number of *drivers* that allow the system to interact with specific items of hardware such as displays and printers.

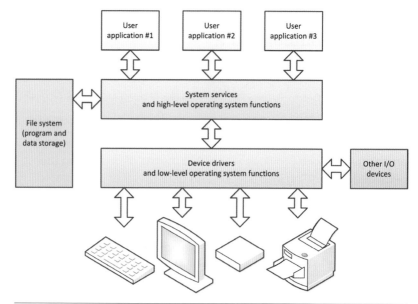

Figure 5.5 How a computer's operating system helps to manage the system's resources and acts as an interface between users and system hardware.

Test your knowledge 5.6

Explain **three** functions of a computer operating system.

Test your knowledge 5.7

Name **three** different computer operating systems.

Activity 5.2

Find out which operating system and which version is used in the computer system that you've investigated in Activity 5.1.

Learning outcome 5.3

Identify the application of standard software packages

Engineers often use the same standard software packages that are used by any modern business. These packages include:

- word-processing software such as Microsoft Word and Corel WordPerfect
- spreadsheet packages such as Microsoft Excel and OpenOffice Calc
- presentation software such as Microsoft PowerPoint and Apple Keynote
- database applications such Microsoft Access and FileMaker Pro
- drawing packages such as Microsoft Visio and Corel DRAW
- photo-editing packages like Adobe Photoshop and Serif PhotoPlus
- desktop publishers such as Microsoft Publisher and Serif PagePlus
- accounting software such as Sage Accounts and Intuit QuickBooks
- project management software such as Microsoft Project
- Web browsers such as Firefox, Chrome and Internet Explorer.

Several of these popular applications can be bundled with larger packages such as Microsoft Office and Apache OpenOffice. These act as a tightly integrated suite of application programs that all share a common user interface and file system. In an integrated software suite (such as Adobe InDesign, Microsoft Office or Star Office) the output of one program (such as a spreadsheet or graphics program) can be readily imported into or embedded in another program (such as a word processor or page setting program).

Software packages allow you to perform everyday tasks such as writing a letter, editing an image or sending an e-mail message. They work through the operating system to access the underlying hardware. So, for example, when you have finished writing a letter the word processor will send a request to the operating system that creates a file from your text, saving it to the hard disk so that it can be recalled at some later time. When the time comes to print the letter, the word processor will pass the text file back through the operating system so that it can be printed.

Key point

Applications software packages allow you to perform everyday tasks such as writing a letter, editing an image or sending an e-mail message. Applications software is often bundled into software suites that provide tight integration between particular applications such as word processors, spreadsheets and database packages.

Key point

Operating system software makes it possible for application software packages to access the underlying hardware of a computer system. This makes it possible for a particular software package to run on a wide variety of different hardware, including tablets, laptops and desktop computers.

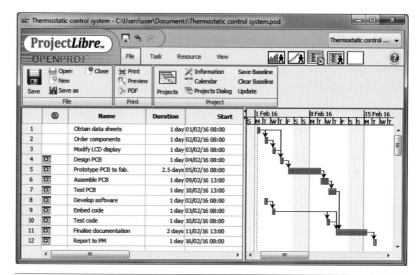

Figure 5.6 Project management software is widely used in engineering to plan and keep track of progress.

Key point

Operating system software makes it possible for application software packages to access the underlying hardware of a computer system. This makes it possible for a particular software package to run on a wide variety of different hardware, including tablets, laptops and desktop computers.

Test your knowledge 5.8

Name **three** different types of software application. Explain briefly what each one is used for.

Utility and bespoke software

In addition to major applications packages computers may also use a number of smaller utility programs to perform specific tasks such as virus and malware protection, file backup and communication. Finally, where there's no existing software that will meet a particular requirement (such as controlling a manufacturing plant) it may be necessary to use bespoke software. Bespoke software is custom software written to perform a particular task. However, since this can often be expensive and time consuming, it is often avoided wherever possible.

Bespoke applications are written using *programming languages* such as C++ and Java. These languages are compiled into code and data that the computer will understand and execute. C++ and Java are sometimes referred to as high-level languages because they can be easily read by humans. Unless you have specialist programming knowledge, low-level languages (such as machine code or assembly language) can be extremely difficult to understand, but they offer fast execution coupled with high efficiency in terms of the use of memory space.

Activity 5.3

In a spreadsheet, data is organized in rows and columns, which collectively are called a worksheet. The intersection of a row and column, called a cell, can contain a label (text), a value (number), or a formula that will allow you to perform a calculation on the data and display the result. Investigate a spreadsheet package such as Microsoft Excel and give a typical example of how a worksheet could be used in engineering.

Activity 5.4

Database software allows you to create and manage a collection of related information, such as stock code, quantity, finish etc. A database query allows you to access and retrieve the data according to a set of criteria that you specify. Investigate a database package such as Microsoft Access and give a typical example of how a database query could be used in engineering.

Learning outcome 5.4

Identify the use of IT/ICT as an information source

Engineers use information derived from a large number of sources. Earlier in Chapter 4 we looked at how written and graphical sources can be used to communicate information. In this section we look at how IT and ICT is used to provide a highly effective and efficient source of information that can be used in a wide variety of engineering contexts. Because of this, many engineering companies are increasingly moving from paper-based information to electronic information that can be updated easily and disseminated widely.

The Internet and the World Wide Web

Although the terms Web and Internet are often used synonymously, they are actually two different things. The Internet is the global association of computers that carries data and makes the exchange of information possible. The World Wide Web is a subset of the Internet – a collection of inter-linked documents that work together using a specific Internet protocol called *hypertext transfer protocol*

(HTTP). In other words, the Internet exists independently of the World Wide Web, but the World Wide Web can't exist without the Internet.

The World Wide Web began in March 1989, when Tim Berners-Lee of the European Particle Physics Laboratory at CERN (the European Centre for Nuclear Research) proposed the project as a means to better communicate research ideas among members of a widespread organization.

Websites are made up of collections of Web pages. Web pages are written in hypertext markup language (HTML), which tells a Web browser (such as Microsoft's Internet Explorer or Mozilla's Firefox) how to display the various elements of a Web page. Just by clicking on a hyperlink, you can be transported to a site on the other side of the world.

A set of unique addresses is used to distinguish the individual sites on the World Wide Web. An *Internet Protocol* (IP) address is a number that identifies a specific computer connected to the Internet. The digits are organized in four groups of numbers (which can range from 0 to 255) separated by full stops. Depending on how an Internet Service Provider (ISP) assigns IP addresses, you may have one address all the time or a different address each time you connect.

Every Web page on the Internet, and even the objects that you see displayed on Web pages, has its own unique address, known as a *uniform resource locator* (URL). The URL tells a browser exactly where to go to find the page or object that it has to display.

Figure 5.7 An example of an engineering company's website displayed in a Web browser.

Test your knowledge 5.9

What do the following initials stand for?

a) HTTP

b) HTML

c) IP

d) ISP

e) URL

Search engines

Being able to locate the information that you need from a vast number of sites scattered across the globe can be a daunting prospect. However, since this is a fairly common requirement, a special type of site, known as a search engine, is available to help you with this task.

Search engines such as Google or Yahoo Search use automated software called Web crawlers or spiders. These programs move from website to website, logging each site title, URL and at least some of its text content. The object is to hit millions of websites and to stay as current with them as possible. The result is a long list of websites placed in a database that users search by typing in a keyword or phrase.

Key point

The Internet provides a means of accessing a vast amount of information. Search engines, search sites and Web directories can usually help you locate the information that you need.

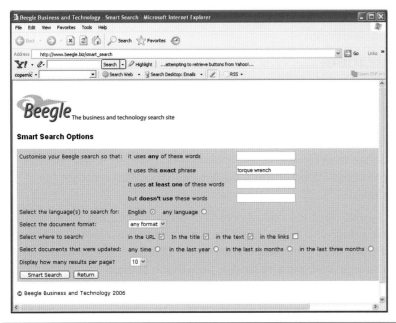

Figure 5.8 A typical search engine used to search for the term 'torque wrench'. Note that the advanced options provide you with a way of narrowing your search to an exact phrase.

Figure 5.9 A Web-based reference source being used to locate and display information on fire extinguishers (see Activity 5.6).

Test your knowledge 5.10

What type of application software is required in order to display a Web page?

Intranets

Intranets work like the Web (with browsers, Web servers, and websites) but companies and other organizations use them internally. Companies use them because they let employees share corporate data, but they're cheaper and easier to manage than most private networks—no one needs any software more complicated or more expensive than a Web browser, for instance.

They also have the added benefit of giving employees access to the Web. Intranets are closed off from the rest of the Internet by firewall software, which lets employees surf the Web but keeps all the data on internal Web servers hidden from those outside the company.

> **Key point**
>
> Intranets are accessible within a company, providing information that is immediately accessible to employees. Extranets allow businesses to share information with customers and suppliers. They are generally less secure than intranets.

Test your knowledge 5.11

Explain the difference between an intranet and the Internet.

Test your knowledge 5.12

What application software is required in order to display a
Web page?

Activity 5.5

The motorway bridge at Millau in France is the world's
highest bridge. The engineering company that supplied the
hydraulic system for lifting the temporary piers and pushing
the bridge decks into position was Enerpac. Visit the company's
website at www.enerpac.com and locate information on
the Millau Viaduct project. Use this to answer the following
questions:

1 What is the height and overall length of the bridge?
2 What valley does the bridge cross?
3 Who designed the bridge?
4 How many bridge piers were constructed?
5 How much concrete was used to build the bridge?
6 How was the bridge deck moved into position?
7 What bus system was used by the electronic control systems?

Figure 5.10 A temporary pillar used to support the Millau bridge deck during
construction (see Activity 5.5). Courtesy of Enerpac.

Activity 5.6

Visit the Wikipedia website at www.wikipedia.com and view the entry on 'Fire Extinguishers' (similar to that shown in Figure 5.9). Read the information and use it (together with information from other Web-based resources) to answer the following questions:

1 Who invented the modern fire extinguisher?
2 List the main parts of a fire extinguisher and say what each part is used for.
3 What are the two main types of fire extinguisher bottle?
4 What is a Class-B fire?
5 What European Standard relates to different classes of fire?
6 What name is used in the USA to describe a 'dry powder' fire extinguisher?
7 Why are halon fire extinguishers illegal in the UK?
8 What precaution should be observed when discharging a CO_2 fire extinguisher and why is this necessary?
9 In the UK, how often should a CO_2 fire extinguisher be pressure tested?
10 Why do Class-D fires need special types of fire extinguisher?

Activity 5.7

Many maintenance and repair tasks on a vehicle require the use of a torque wrench. Use information sources (such as a library or Google search) to find out what a torque wrench is and then write a short word-processed technical report (not more than one page of A4) describing a torque wrench and explaining how it is used. Include a diagram or image in your answer.

Activity 5.8

Visit the Draper Tools website at www.drapertools.com and use it to download the instruction manual for a 14.4V Cordless Drill (Part No. CD144V). Use the information obtained from the instruction manual to answer the following questions:

1 What is the no-load speed range?
2 What is the chuck size?
3 What is the drilling capacity when used with mild steel?
4 What is the weight of the drill plus battery?

5 What is the spindle thread for the chuck?

6 What is the cell rating?

7 What is the procedure for battery disposal and why is it important to follow this recommendation?

8 What precaution should be observed when charging the battery and why is this important?

Learning outcome 5.5

Identify the use and application of software packages within engineering

Software applications are widely used in all branches of engineering. You have already met some standard software packages but there are many packages designed specifically for use in particular engineering tasks such as design, testing and simulation. We usually refer to this general class of software as software for *computer-aided engineering* (CAE). Computer-aided engineering is about automating the various stages that go into providing an engineered product or service. When applied effectively, CAE ties all of the functions within an engineering company together. Within a true CAE environment, information (i.e. data) is passed from one computer-aided process to another. This may involve computer simulation, computer-aided drawing (CAD) and computer-aided manufacture (CAM).

The term CAD/CAM is used to describe the integration of computer-aided design, drafting and manufacture. Another term, CIM (computer-integrated manufacturing), is often applied to an environment in which computers are used as a common link that binds together the various different stages of manufacturing a product, from initial design and drawing to final product testing.

While all of these abbreviations can be confusing (particularly as some of them are often used interchangeably), it is worth remembering that 'computer' appears in all of them. What we are really talking about is the application of computers within engineering. Nowadays, the boundaries between the strict disciplines of CAD and CAM are becoming increasingly blurred and fully integrated CAE systems are becoming commonplace in engineering companies.

In Chapter 4 we've already introduced CAD as a means of producing engineering drawings. Several different types of drawing are used in engineering. An example of a typical CAD drawing is shown in Figure 5.11.

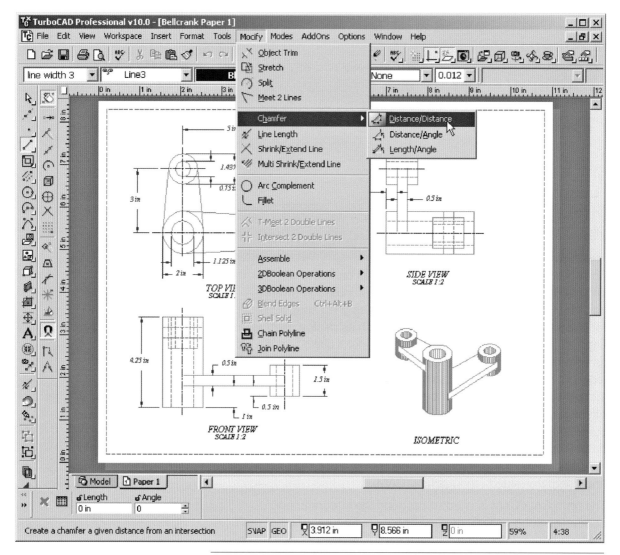

Figure 5.11 Drop-down menus and extensive toolbars are available in this powerful CAD package which incorporates both 2D and 3D features.

Computer-aided manufacture (CAM)

Computer-aided manufacture (CAM) encompasses a number of more specialized applications of computers in engineering including computer integrated manufacturing (CIM), manufacturing system modelling and simulation, systems integration, artificial intelligence applications in manufacturing control, CAD/CAM, robotics and metrology.

In a modern engineering company, all of the machine tools within a particular manufacturing company may be directly linked to the CAE network through the use of centrally located floor managers which monitor machining operations and provide sufficient memory for complete machining runs.

Manufacturing industries rely heavily on computer-controlled manufacturing systems. Some of the most advanced automated systems are employed by those industries that process petrol, gas, iron and steel. The manufacture of cars and trucks frequently involves computer-controlled robot devices. Industrial robots are used in a huge range of applications that involve assembly or manipulation of components.

The introduction of CAD/CAM has significantly increased productivity and reduced the time required to develop new products. When using a CAD/CAM system, an engineer develops the design of a component directly on the display screen of a computer. Information about the component and how it is to be manufactured is then passed from computer to computer within the CAD/CAM system. After the design has been tested and approved, the CAD/CAM system prepares sequences of instructions for computer numerically controlled (CNC) machine tools and places orders for the required materials and any additional parts (such as nuts, bolts or adhesives). The CAD/CAM system allows an engineer (or, more likely, a team of engineers) to perform all the activities of engineering design by interacting with a computer system (invariably networked) before actually manufacturing the component in question using one or more CNC machines linked to the CAE system.

Key point

CAD/CAM can significantly increase productivity and reduce the time required to develop a new product. When using a CAD/CAM system, an engineer develops the design of a component directly on the display screen of a computer. When the design is complete, relevant data is passed to the computers responsible for actually manufacturing the component.

Figure 5.13 Some typical components manufactured using CNC processes and steel bar.

Figure 5.12 Engineers programming a CNC machine.

Test your knowledge 5.13

Explain the advantages of using CAD/CAM in the design and manufacture of an engineered product.

Figure 5.14 Some typical components manufactured using CNC processes and cast alloy components.

Test your knowledge 5.14

Explain the benefits of CAM compared with manual manufacturing techniques and give **three** examples of its use.

2D CAD

Two-dimensional (2D) CAD packages are widely used by engineers for creating conventional 'flat' drawings. Typical of these 2D packages are 'industry standard' packages like Autosketch, AutoCAD, TurboCAD, DesignCAD and LibreCAD (note that several of these packages also support 3D design).

Older 2D CAD programs accept text commands or a combination of buttons followed by coordinates or other parameters entered as text. All modern packages make more use of the graphical user interface (GUI) available in modern computers but they may still require precise dimensions to be entered from the keyboard during a drawing session. For example, the following steps can be used to construct the LibreCAD drawing shown in Figure 5.15:

1 Click in the tool window to open the 'Circle' tool window.
2 Select the 'Centre, Point' tool.
3 Enter the coordinates (0,0) of the centre of the circle.
4 Specify a radius of 50mm and the circle will be drawn with a diameter of 100mm.
5 Click in the tool window to open the 'Line' tool window.

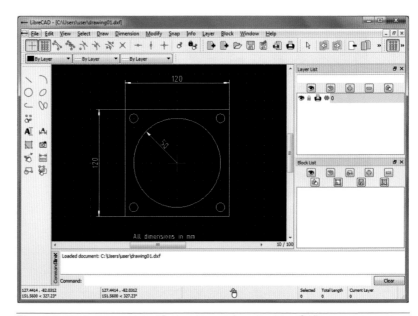

Figure 5.15 An incomplete 2D drawing made using LibreCAD.

6 Select the 'Rectangle' tool.

7 Enter -60, -60 as the coordinates of the first corner.

8 Enter 60, 60 as the coordinates of the opposite corner and the rectangle will be drawn with a side length of 120mm.

Dimensions can be added by selecting 'Dimensions' from the menu bar and then 'Horizontal' or 'Vertical' and then clicking on the respective lines. The mouse can be used to drag the dimension lines away from the drawing in order to separate them by an appropriate distance.

The four mounting holes can then be drawn by adding four more circles with a radius of 5mm centred on (–50, –50), (–50, 50), (50, –50) and (50, 50).

Most modern CAD programs are reasonably intuitive and will allow you to quickly create drawings using standard templates and symbol libraries. A variety of drawing tools will be provided which will allow you to assign different properties to drawing entities (such as a dashed line or a hatched rectangle) or to snap, glue or group entities together. Intelligent connectors (which rebuild automatically when you reposition objects and avoid crossing other objects) allow you to easily create charts and schematics.

If you've never used a CAD package, you may find that it takes some time to become proficient with it – so don't be too disappointed if your first efforts don't look too professional! With time (and a little effort) you should be able to produce engineering drawings to an acceptable standard.

The features available from a modern 2D CAD package include:

- various modes for creating lines, arcs, circles, ellipses, parallel line, angle bisectors, etc.
- support (import and export) for standard CAD file formats (e.g. DXF)
- standard and user-defined symbol libraries
- text in different fonts and sizes
- automatic dimensioning of distances, angles, diameters and tolerances
- solid fills and hatching
- support for layers and blocks (which can be manipulated as a separate entity)
- selection and modification tools (move, rotate, mirror, trim, stretch etc.)
- snapping to grid and to objects (endpoints, centres, intersections etc.)
- multiple undo and redo levels

- support for various units including metric, imperial, degrees, radians etc.
- import and export of bitmaps and other images in various format (e.g. BMP, JPEG, PNG etc.)
- OLE (Object Linking and Embedding) compatibility (this means that a CAD drawing can be inserted and edited within any other OLE-compatible Windows application).

If you would like to practise some CAD skills at an early stage in this unit (and to help you get acquainted with the commands and techniques used in CAD packages), you will find several simple 2D CAD packages that can be downloaded freely from the Web. Most of these support industry standard DWG and DXF formats and use basic commands broadly compatible with those used in several of the more powerful CAD packages.

Activity 5.9

Visit www.librecad.org and download you own copy of the Open Source LibreCAD 2D drawing package. Use it to complete the drawing shown in Figure 5.15 (note that several dimensions are missing from the drawing). Print a copy of the finished drawing.

3D CAD

A 3D CAD package produces a model of a component part or product which can be viewed from any desired angle. 3D CAD packages may operate in *wire-frame mode* (in which you will see a skeletal outline of the component or product, as shown in Figure 5.16) or in *render mode* (in which you will see a more realistic shaded image, see Figures 5.20 and 5.21). CAD packages will usually provide you with both *orthographic views* (see Figure 5.17) and *isometric views* (as shown in Figure 5.18). These views will often be displayed in multiple windows so that it is possible to view a model from several different points at the same time. In addition, more advanced packages provide a *camera view* that can be used to provide additional perspective and which will allow you to examine a component part or a product from any desired angle.

3D CAD packages use a three-axis coordinate system and a plane of reference (the *workplane*) as shown in Figure 5.19. In addition, some packages also allow you to define your own user coordinate system which has a different workplane that travels with an object and which is usually defined first in 2D mode.

The workplane is the plane in which a 2D object is initially created (i.e. as a 2D drawing). In 2D mode, you will normally do all your

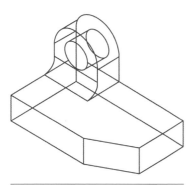

Figure 5.16 A 3D wire-frame view of a component.

work in the same workplane but in 3D mode it is frequently necessary to change the workplane in order to perform all the required commands.

One of the most important tools for controlling the view of a 3D model is that which allows you to render an object so that it looks both solid and realistic. In normal *rendering* modes, all 3D objects are displayed as shaded (with or without hidden lines). Higher-level rendering will enable you to view materials and textures, thus providing a more realistic image of what a component or your model will actually look like.

In order to create a realistic rendered view, lighting effects must be added and the positions of the light sources are usually made adjustable. In high-end packages, it may also be possible to further

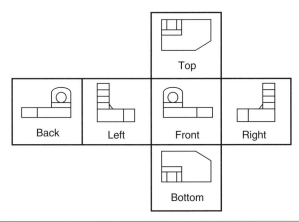

Top

Back   Left   Front   Right

Bottom

Figure 5.17 Orthographic views (see later) of the component shown in Figure 5.16.

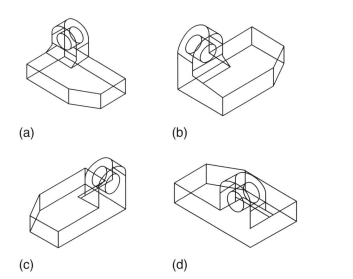

(a)   (b)

(c)   (d)

Figure 5.18 Isometric views (see later) of the object shown in Figure 5.16.

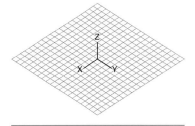

Figure 5.19 A 3D workplane showing reference axes.

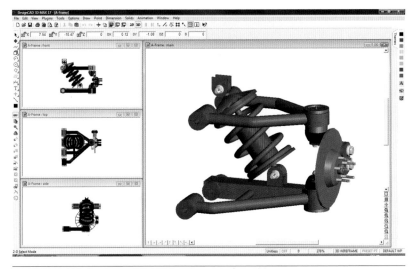

Figure 5.20 A 2D and 3D rendered model of a vehicle component.

enhance the rendering of the image by assigning materials and luminance qualities to objects. This adds further realism to the final generated image.

Test your knowledge 5.15

Sketch a 3D workplane and label the axes.

Other software applications

Many other specialized software applications are used in different branches of engineering, including dedicated plant and process control applications. In Figure 5.21, a computer is used to control a manufacturing process which involves the fermentation of a liquid at a particular temperature. Since this is a continuous process the control system is designed to maintain the level of fluid in the tank by opening and closing various valves.

Software applications are widely used in the design and manufacture of electronic equipment. Figure 5.22 shows how an electronic circuit can be designed and tested prior to manufacture. The software uses a tool called SPICE (Simulation Program with Integrated Circuit Emphasis), which checks the integrity of a circuit design and predicts its behaviour. Software is also widely used to design and manufacture printed circuits on which electronic components are mounted. Figure 5.23 shows the layout of a circuit with track links automatically routed using software. When layout and routing is complete, the board design can then be sent to CNC machines for production equipment.

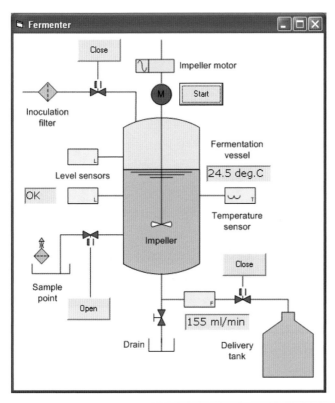

Figure 5.21 Software is widely used in dedicated plant and process control applications.

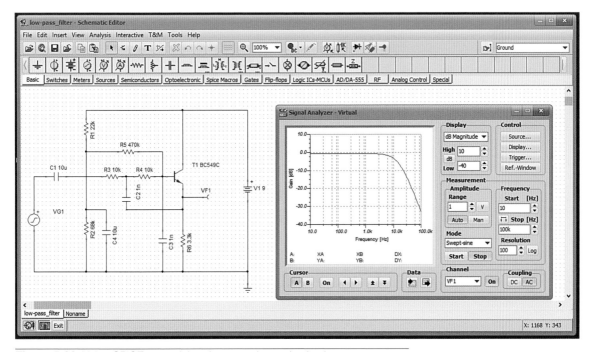

Figure 5.22 Using SPICE to model and test an electronic circuit.

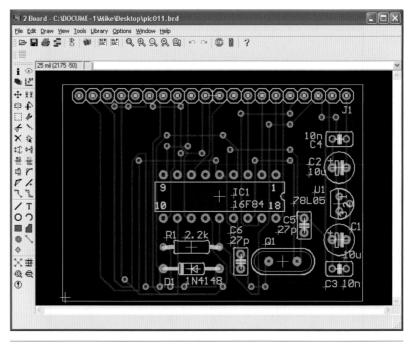

Figure 5.23 Printed circuit design is an ideal task for software.

Test your knowledge 5.16

Explain why a 3D CAD model is rendered prior to viewing.

Test your knowledge 5.17

Why is it necessary to be able to select and adjust the light source when rendering a 3D CAD object?

Activity 5.10

Investigate **one** specialized software application used in engineering. In what branch (or branches) of engineering is this application designed for use? Write a brief description of the application and explain what it does.

Review questions

1. Identify **three** types of IT hardware used in engineering applications. Explain the purpose of each one.

2. Explain why a computer needs an operating system. How does this differ from an application program?

3. Identify **three** standard software application packages and say what each one is used for.

4. List **three** information sources accessible through IT/ICT. Give an example of the use of each one.

5. Explain the difference between 2D and 3D CAD.

6. State **three** advantages of using CNC when compared with manual methods of engineering manufacture.

7. Describe **one** specialized software application used in engineering.

Chapter checklist

Learning outcome	Page number
5.1 Identify the types of IT and ICT hardware used in engineering applications.	96
5.2 Identify the types of IT and ICT operating systems used in engineering applications.	101
5.3 Identify the application of standard software packages.	103
5.4 Identify the use of IT/ICT as an information source.	105
5.5 Identify the use and application of software packages within engineering.	111

CHAPTER 6

Tools and techniques

Learning outcomes

When you have completed this chapter you should understand the basic tools and techniques used in engineering, including being able to:

6.1 Identify work- and tool-holding methods and applications.

6.2 Identify cutting tool types and their uses.

6.3 Identify drill bits and their uses.

6.4 Identify the basic screw thread forms and their uses.

6.5 Identify the basic methods of work assembly.

6.6 Identify basic fault-finding techniques to simple problems.

Chapter summary

This chapter introduces you to common tools and techniques used in engineering. As an engineer you will need to develop proficiency with a number of skills, including cutting, forming, joining and assembling parts and components. In order to be able to do this, you will need to understand the various processes involved, such as how to select and use a drill, how to make a soldered joint and how to select and fit threaded components such as nuts, bolts and screws. All of this is something that will take you into a practical engineering workshop environment. What you learn and practise there will undoubtedly form the foundation of your 'skill set' as an engineer.

Learning outcome 6.1

Identify work- and tool-holding methods and applications

Being able to hold a workpiece accurately, securely and safely brings a number of benefits, not the least of which is that it helps to improve precision, reduce errors and minimize waste. At the most basic level, work-holding can be based on the use of simple bench clamps and jigs. At the most sophisticated level it can involve vacuum systems that are highly effective for use with thin or irregular-shaped parts. Later in this chapter we will introduce work-holding arrangements that can be used with machine tools, such as pillar drills, lathes and milling machines but, in this section we will start by looking at the most basic method of work-holding, the engineer's or fitter's bench vice.

Bench vice

A bench vice should be substantial and rigid. This is essential if accurate work is to be performed on it. The vice should be positioned so that it is well lit by both natural and artificial light without glare or shadows. A plain screw vice is shown in Figure 6.1a and a quick action vice is shown in Figure 6.1b. In the latter type of vice, the jaws can be quickly pulled apart when the lever at the side of the screw handle is released. The screw is used for closing the jaws and clamping the work in the usual way.

The jaws of an engineer's vice are serrated and hardened to prevent the work from slipping. Unfortunately, this is also liable to damage the surface of the work. For fine work with finished surfaces, the serrated jaws should be replaced with hardened and ground

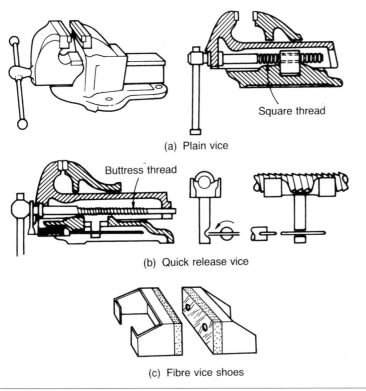

(a) Plain vice

Square thread

Buttress thread

(b) Quick release vice

(c) Fibre vice shoes

Figure 6.1 An engineer's vice.

smooth jaws. Alternatively, vice shoes can be used. These are faced with a fibre compound and can be slipped over the serrated jaws when required. A pair of typical vice shoes are shown in Figure 6.1c. In the sections that follow we will be looking at methods of work-holding appropriate for use with machine tools such as lathes and milling machines.

Learning outcome 6.2

Identify cutting tool types and their uses

Cutting tools

Before we can discuss the cutting tools we use for bench work and 'fitting' we need to look at the way metal is cut. Here are the basic facts.

Wedge angle

If you look at a hacksaw blade, as shown in Figure 6.2a, you can see that the teeth are wedge shaped. Figure 6.2b shows how the wedge angle increases as the material gets harder. This strengthens

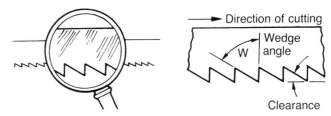

(a) Hacksaw blade showing the wedge angle

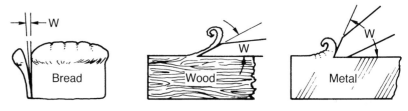

(b) Wedge angles for various materials

Figure 6.2 Hacksaw blade and wedge angles.

the cutting edge and increases the life of the tool. At the same time it reduces its ability to cut. Try cutting a slice of bread with a cold chisel!

Clearance angle

If you look at the hacksaw blade in Figure 6.2a, you can see that there is a clearance angle behind the cutting edge of the tooth. This is to enable the tooth to cut into the work.

Rake angle

This angle controls the cutting action of the tool. It is shown in Figure 6.3. I hope you can see that the wedge angle, clearance

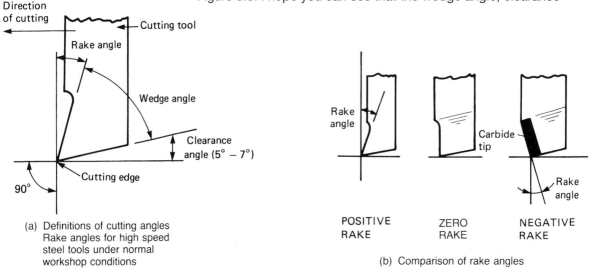

(a) Definitions of cutting angles
Rake angles for high speed steel tools under normal workshop conditions

(b) Comparison of rake angles

Figure 6.3 Rake angle.

angle and rake angle always add up to 90°. This is true even when the rake angle is zero or negative as shown in Figure 6.3b. The greater the rake angle the more easily the tool will cut. Unfortunately, the greater the rake angle, the smaller the wedge angle will be and the weaker the tool will be. Therefore the wedge and rake angles have to be a compromise between ease of cutting and tool strength and life. The clearance angle remains constant at between 5° and 7°.

Table 6.1 Rake angle for various materials.

Material	Rake angle
Aluminium alloy	30°
Brass (ductile)	14°
Brass (free-cutting)	0°
Cast iron	0°
Copper	20°
Phosphor bronze	8°
Mild steel	25°
Medium carbon steel	15°

Test your knowledge 6.1

Calculate the wedge angle for a tool if the clearance angle is 5°and the rake angle is 17°.

Test your knowledge 6.2

Explain why vice shoes are essential when carrying out vice work on a soft material.

Test your knowledge 6.3

Explain why metal cutting tools have a larger wedge angle than wood cutting tools.

Test your knowledge 6.4

Explain why metal cutting tools need a clearance angle.

Orthogonal and oblique cutting

Figure 6.4a is a pictorial representation of the single point cutting tool shown in Figure 6.2. Notice how the cutting edge is at right angles to the direction in which the tool is travelling along the work. This is called orthogonal cutting. Now look at Figure 6.4b. Notice how the cutting edge is inclined at an angle to the direction of cut. This is called oblique cutting. Oblique cutting results in a better finish than orthogonal cutting, mainly because the chip is thinner for a given rate of metal removal. This reduced thickness and the geometry of the tool allows the chip to coil up easily in a spiral.

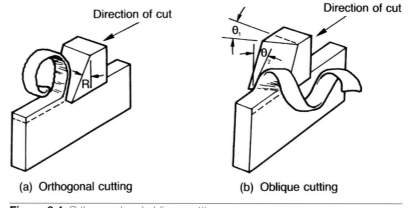

(a) Orthogonal cutting   (b) Oblique cutting

Figure 6.4 Orthogonal and oblique cutting.

Test your knowledge 6.5

Sketch an example of a tool that cuts orthogonally and a tool that cuts obliquely.

Coolant

Apart from threading operations, it is very rare to use a coolant or lubricant when using hand tools. However, the conditions are very different when using machine tools. Large amounts of metal are removed quickly, considerable energy is used to do this, and this energy is largely converted into heat at the cutting zone. The rapid temperature rise of the work and the cutting tool can lead to inaccuracy and short tool life. A *coolant* is required to prevent this. The chip flowing over the rake face of the tool results in wear. A lubricant is required to prevent this. Usually, coolants are poor lubricants, and lubricants are poor coolants. For general machining an emulsion of cutting oil (which also contains an *emulsifier*) and water is used. This has a milky-white appearance and is commonly known as 'suds'.

Files

Files are the most widely used and important tools for the fitter. The main parts of a file are named in Figure 6.5a. Files are forged to shape from 1.2% plain carbon steel. After forging, the teeth are machine cut by a chisel shaped tool, as shown in Figure 6.5b. The teeth of a single cut are wedge shaped with the rake and clearance angles essential for metal cutting. Most files used in general engineering are double-cut. That is they have two rows of cuts at an angle to each other, as shown in Figure 6.5c.

Files are classified by the following features:

- length
- kind of cut
- grade of cut (roughness)
- profile
- cross-sectional shape or most common use.

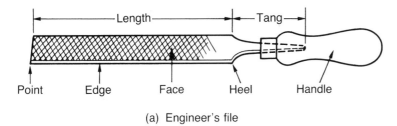

(a) Engineer's file

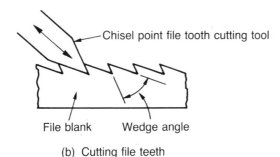

(b) Cutting file teeth

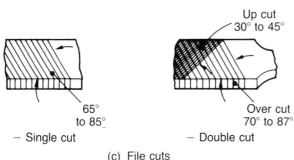

(c) File cuts

Figure 6.5 An engineer's file.

The grades of cut are: rough, bastard, second, smooth and dead smooth. These cuts vary with the length of a file. For example a short, second cut file will be smoother than a longer, smooth file. The profiles and cross-sectional shapes of some typical files are shown in Figure 6.6.

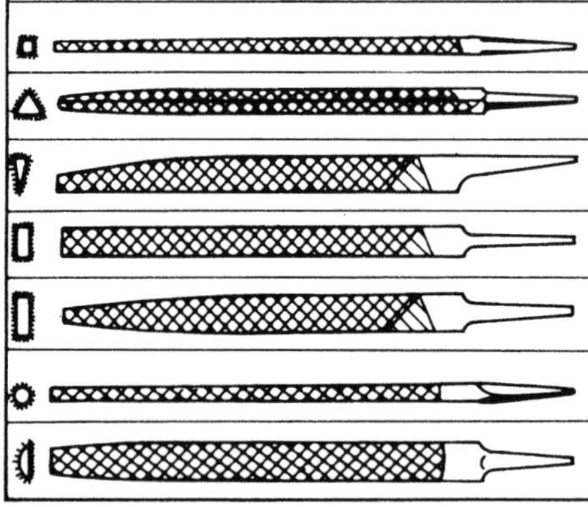

Type of file		Applications
	Square	Filing of keyways and slots
	Three square	Filing of angled surfaces
	Knife	Filing of acute angles
	Hand	These two files are the general-purpose tools for filing flat surfaces and convex profiles
	Flat	
	Round	Used for enlarging or elongating holes
	Half round	Filing of concave profiles

Figure 6.6 Types of file and applications.

Figure 6.7 shows how a file should be held and used. To file flat is very difficult and the skill only comes with years of continual practice. *Cross filing* is used for rapid material removal. *Draw filing* is only a finishing operation to improve the surface finish. It removes less metal per stroke than cross filing and can produce a hollow surface, unless care is taken.

The spaces between the teeth of a file tend to become clogged with bits of metal. This happens mostly when filing soft metals. It is called *pinning*. The clogged teeth tend to leave heavy score marks in the surface of the work. These marks are difficult to remove. The file should be kept clean and a little chalk should be rubbed into the teeth to prevent pinning. Files are cleaned with a file brush called a *file card*.

Test your knowledge 6.6

Explain why cutting fluids (suds) do not have to be used when filing, but are used when turning.

Test your knowledge 6.7

Explain what is meant by a file having a 'safe edge' and name the more common types of file that have a 'safe edge'.

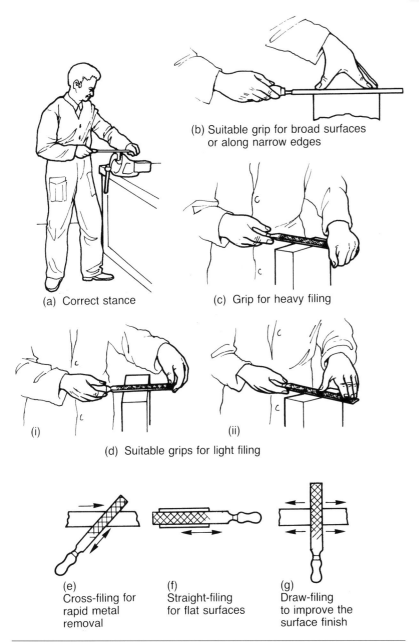

(a) Correct stance

(b) Suitable grip for broad surfaces or along narrow edges

(c) Grip for heavy filing

(i)

(ii)

(d) Suitable grips for light filing

(e) Cross-filing for rapid metal removal

(f) Straight-filing for flat surfaces

(g) Draw-filing to improve the surface finish

Figure 6.7 Correct use of a file.

Test your knowledge 6.8

Explain why it is necessary to clean the teeth of a file.

Test your knowledge 6.9

Explain the difference between cross filing and draw filing.

Hacksaws and sawing

A typical hacksaw frame and blade is shown in Figure 6.8a. The frame is adjustable so that it can be used with blades of various lengths. It is also designed to hold the blade in tension when the wing nut is tightened. The blade is put into the frame so that the teeth cut on the forward stroke. Figure 6.8b shows how a hacksaw should be held when being used.

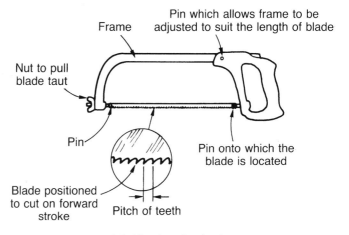

(a) Metal cutting hacksaw

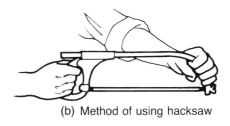

(b) Method of using hacksaw

Figure 6.8 A typical hacksaw frame and blade.

A variety of different hacksaw blade types is available:

- High-speed steel 'all hard' blades are the most rigid and give the most accurate cut. Unfortunately, they are brittle and easily broken when used by an inexperienced person.
- High-speed steel 'soft back' blades have a good life and, being more flexible, are less easily broken.
- Carbon steel flexible blades are satisfactory for occasional use on soft non-ferrous metals. They are cheap and not easily broken. Unfortunately, they only have a limited life when cutting steels.

To prevent the blade from jamming in the slot it makes as it cuts, all saw blades are given a *set*. This is shown in Figure 6.9. Coarse pitch hacksaw blades and power-saw blades have the

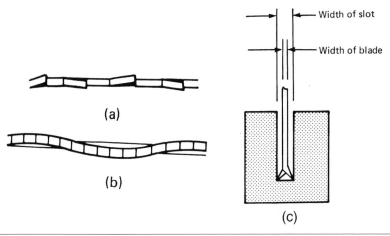

Figure 6.9 The set of a hacksaw blade.

individual teeth set to the left and to the right with either the intermediate teeth or every third tooth left straight to clear the slot. This is shown in Figure 6.9a. Unfortunately, this is not possible with fine pitch blades so the blade as a whole is given a *wave set* as shown in Figure 6.9b. The effect of set on the cut being made is shown in Figure 6.9c. If you have to change a blade part way through a cut, never continue in the old slot. Because the set of the old blade will have worn, the new blade will jam in the old cut and break. Always start a new cut to the side of the failed cut.

The size of a hacksaw blade is specified in terms of the length (between the fixing hole centres), the width and the thickness. Blades are generally supplied in two lengths, 250mm and 300mm. Somewhat confusingly, blades are sometimes also described by the number of teeth per inch (TPI). Blades can have have 14, 18, 24, 32 teeth per inch (approx. 25mm).

The fewer teeth per inch the coarser will be the cut, the more quickly will the metal be removed, and the greater will be the set so that there is less chance of the blade jamming. However, there should always be at least three teeth in contact with the work at any one time. Therefore, the thinner the metal being cut the finer the pitch of the blade that should be used. Some typical examples are given in Table 6.2.

Screw thread cutting

Internal screw threads are cut with taps. A set of straight fluted hand taps are shown in Figure 6.10. The difference between them is the length of the lead. The taper tap should be used first to start the thread. Great care must be taken to ensure that the tap is upright in the hole and it should be checked with a try square. The second tap is used to increase the length of thread and can be used for

Table 6.2 Hacksaw blade sizes and applications.

Teeth per inch	Material to be cut	Blade applications
32	Up to 3mm	Thin sheets and tubes Hard and soft materials (thin sections)
24	3mm to 6mm	Thicker sheets and tubes Hard and soft materials (thicker sections)
18	6mm to 12mm	Heavier sections such as mild steel, cast iron, aluminium, brass, copper, bronze
14	Greater than 12mm	Soft materials (such as aluminium, brass, copper, bronze) with thicker sections

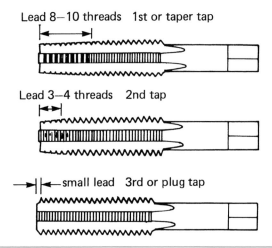

Figure 6.10 Hand taps.

finishing if the tap passes through the work. The third tap is used for *bottoming* in blind holes.

The hole to be threaded is called a *tapping size* hole and it is the same size or only very slightly larger than the core diameter of the thread. Drill diameters for drilling tapping size holes for different screw threads can be found in sets of workshop tables. For example, the tapping size drill for an M10 × 1.5 thread is 8.5mm diameter.

A *tap wrench* is used to rotate the taps. There are a variety of different styles available depending upon the size of the taps. An example of a suitable wrench for small taps is shown in Figure 6.11. Taps are very fragile and are easily broken, particularly in the

Figure 6.11 A tap wrench.

small sizes. Once a tap has been broken into a hole, it is virtually impossible to get it out without damaging or destroying the workpiece.

Taps are relatively expensive and should be looked after carefully. High-speed steel ground thread taps are the most expensive. However, they cut very accurate threads and, with careful use, last a long time. Carbon steel cut thread taps are less accurate and less expensive and have a reasonable life when cutting the softer non-ferrous metals. Whichever sort of taps are used, they should always be well lubricated. Traditionally, tallow was used, but nowadays proprietary screw-cutting lubricants are available that are more effective.

External threads are cut using split button dies in a die holder, as shown in Figure 6.12. One face of the die is always marked up with details of the thread and the maker's logo. This should be visible when the die is in the die-holder. Then the lead is on the correct side for starting the cut. Screw A is used to spread the die for the first cut. The screws marked B are used to close the die until it gives the correct finishing cut. This is judged by using a standard nut or a screw thread gauge. The nut or gauge should run up the thread without binding or without undue looseness.

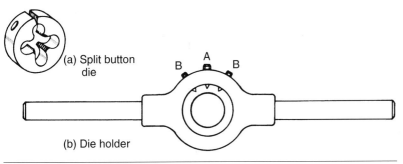

(a) Split button die

(b) Die holder

Figure 6.12 A die holder.

Again, the die must be started square with the workpiece or a 'drunken' thread will result. Also, a thread cutting lubricant should be used. Like thread cutting taps, dies are available in carbon steel cut thread and high-speed steel ground thread types. For both taps and dies, each set only cuts one size and pitch of thread and one thread form.

Spanners and keys

In addition to cutting tools a fitter should also have a selection of spanners and keys available for dismantling and assembly purposes. Figure 6.13 shows a selection of spanners and keys. These are carefully proportioned so that a person of

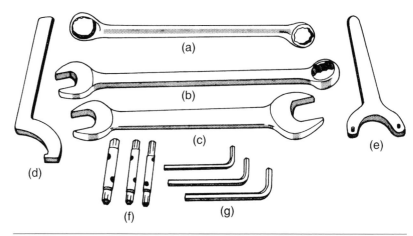

Figure 6.13 A selection of spanners and keys.

average strength will be able to tighten a screwed fastening correctly.

Use of a piece of tubing to extend a spanner or key is very bad practice. It strains the jaws of the spanner so that it becomes loose and may slip. It may even crack the jaws of the spanner so that they break. In both cases this can lead to nasty injuries to your hands and even a serious fall if you are working on a ladder. Also it over-stresses the fastening which will be weakened or even broken. Always check a spanner for damage and correct fit before using it. A torque spanner should be used to tighten important fastenings.

Test your knowledge 6.10

Explain how a button die is adjusted to cut the required diameter in a thread.

Test your knowledge 6.11

With reference to workshop tables, select tapping size drills for the following threads:

a) M8 × 1.0
b) M12 × 1.75
c) 4 BA

Test your knowledge 6.12

Explain why important fastenings are tightened with a torque spanner.

Test your knowledge 6.13

Sketch and name each of the spanners and keys shown in Figure 6.13.

The centre lathe

The main purpose of a centre lathe is to produce external and internal cylindrical and conical (tapered) surfaces. It can also produce plain surfaces and screw threads. Figure 6.14a shows a typical centre lathe.

- The bed is the base of the machine to which all the other sub-assemblies are attached. Slideways accurately machined on its

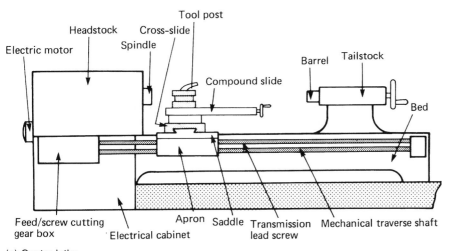

(a) Centre lathe

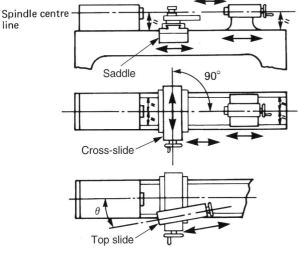

(b) Movements

Figure 6.14 A centre lathe.

top surface provide guidance for the saddle and the tailstock. These slideways also locate the headstock so that the axis of the spindle is parallel with the movement of the saddle and the tailstock. The saddle or carriage of the lathe moves parallel to the spindle axis as shown in Figure 6.14b.

- The cross-slide is mounted on the saddle of the lathe. It moves at 90° to the axis of the spindle, as shown in Figure 6.14c. It provides in-feed for the cutting tool when cylindrically turning. It is also used to produce a plain surface when facing across the end of a bar or component.
- The top-slide (compound-slide) is used to provide in-feed for the tool when facing. It can also be set at an angle to the spindle axis for turning tapers, as shown in Figure 6.14b.

The cutting movements of a centre lathe are summarized in Table 6.3.

Table 6.3 Centre lathe movements.

Cutting movement	Hand or power traverse	Means by which movement is achieved	Turned feature
Tool parallel to the spindle centre line	Both	The saddle moves along the bed slideways	A parallel cylinder
Tool at 90° to the spindle centre line	Both	The cross-slide moves along a slideway machined on the top of the saddle	A flat face square to the spindle centre line
Tool at an angle relative to the spindle centre line	Hand	The compound-slide is rotated and set at the desired angle relative to the centre line	A tapered cone

Work-holding in the lathe

The work to be turned can be held in various ways. We will now consider the more important of these. The centre lathe derives its name from this basic method of work-holding. The general layout is shown in Figure 6.15a. Centre holes are drilled in the ends of the bar and these locate on centres in the headstock spindle and the tailstock barrel. A section through a correctly centred component is shown in Figure 6.15b. The centre hole is cut with a standard centre drill. The main disadvantage of this method of work-holding is that no work can be performed on the end of the component. Work that has been previously bored can be finish turned between centres using a taper mandrel as shown in Figure 6.15c.

Four-jaw chuck

Chucks are mounted directly onto the spindle nose and hold the work securely without the need for a back centre. This allows the

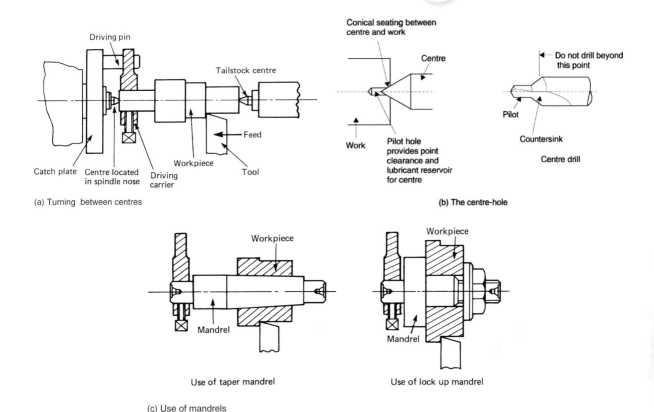

(a) Turning between centres

(b) The centre-hole

Use of taper mandrel

Use of lock up mandrel

(c) Use of mandrels

Figure 6.15 Work-holding in a centre lathe.

end of the work to be faced flat. It also allows for the work to have holes bored into it or through it.

In the four-jaw chuck, the jaws can be moved independently by means of jack-screws. As shown in Figure 6.16a, the jaws can also be reversed and the work held in various ways, as shown in Figure 6.16b. As well as cylindrical work, rectangular work can also be held, as shown in Figure 6.16c. Because the jaws can be moved independently, the work can be set to run concentrically with the spindle axis to a high degree of accuracy. Alternatively, the work can be deliberately set off-centre to produce eccentric components as shown in Figure 6.16d.

Three-jaw chuck

The self-centring, three-jaw chuck is shown in Figure 6.17a. The jaws are set at 120° and are moved in or out simultaneously (at the same time) by a scroll when the key is turned. SAFETY: This key must be removed before starting the lathe or a serious accident can occur. When new and in good condition, this type of chuck can hold cylindrical and hexagonal work concentric with the spindle axis to a high degree of accuracy. In this case the

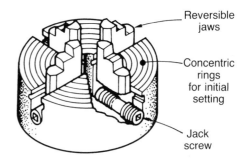

(a) Independent four-jaw chuck

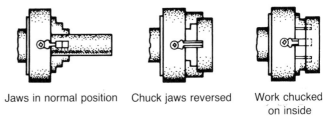

Jaws in normal position   Chuck jaws reversed   Work chucked on inside

(b) Methods of holding work in a chuck

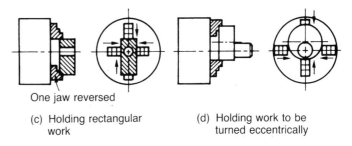

One jaw reversed

(c) Holding rectangular work

(d) Holding work to be turned eccentrically

Figure 6.16 Four-jaw chuck.

jaws are not reversible, so it is provided with separate internal and external jaws. In Figure 6.17a the internal jaws are shown in the chuck, and the external jaws are shown at the side of the chuck. Again, the chuck is mounted directly on the spindle nose of the lathe.

Work to be turned between centres is usually held in a three-jaw chuck while the ends of the bar are faced flat and then centre drilled, as shown in Figure 6.17b.

Face-plate

Figure 6.18 shows a component held on a face-plate so that the hole can be bored perpendicularly to the datum surface. This datum surface is in contact with the face-plate. Note that the face-plate has to be balanced to ensure smooth running. Care must be taken

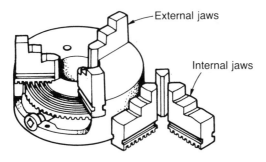

(a) Three jaw self-centring chuck

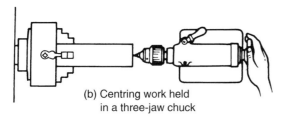

(b) Centring work held
in a three-jaw chuck

Figure 6.17 Three-jaw chuck.

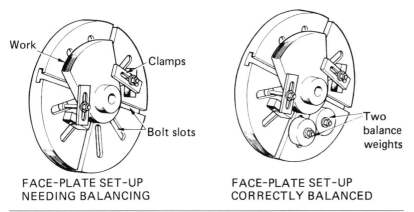

FACE-PLATE SET-UP
NEEDING BALANCING

FACE-PLATE SET-UP
CORRECTLY BALANCED

Figure 6.18 Face-plate set-up.

to check that the clamps will hold the work securely and do not foul the machine. The clamps must not only resist the cutting forces, but they must also prevent the rapidly rotating work from spinning out of the lathe.

Turning tools

Figure 6.19a shows a range of turning tools and some typical applications. Figure 6.19b shows how the metal-cutting wedge also applies to turning tools. Turning tools are fastened into a tool-post which is mounted on the top-slide of the lathe. There are many different types of tool-post. The four-way turret tool-post shown in Figure 6.19c allows four tools to be mounted at any one time.

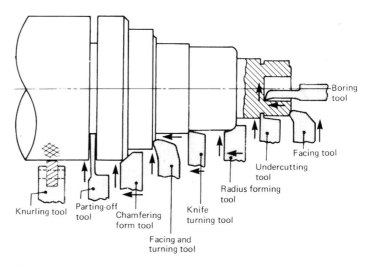

(a) Turning tools

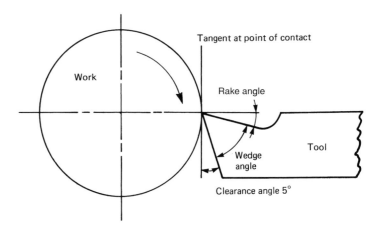

(b) Turning tool angles

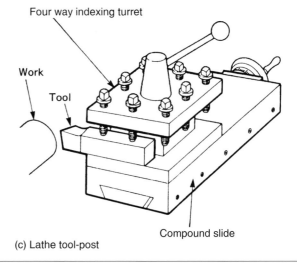

(c) Lathe tool-post

Figure 6.19 Turning tools.

Parallel turning

Figure 6.20a shows a long bar held between centres. To ensure that the work is truly cylindrical with no taper, the axis of the tailstock centre must be in line with the axis of the headstock spindle. The saddle traverse provides movement of the tool parallel with the workpiece axis. You take a test cut and measure the diameter of the bar at both ends. If all is well, the diameter should be constant all along the bar. If not, the lateral movement of the tailstock needs to be adjusted until a constant measurement is obtained. The depth of cut is controlled by micrometer adjustment of the cross-slide.

While facing and centre drilling the end of a long bar, a fixed steady is used. This supports the end of the bar remote from the chuck. A fixed steady is shown in Figure 6.20b.

If the work is long and slender it sometimes tries to kick away from the turning tool or even climb over the tool. To prevent this happening a travelling steady is used. This is bolted to the saddle opposite to the tool, as shown in Figure 6.20c.

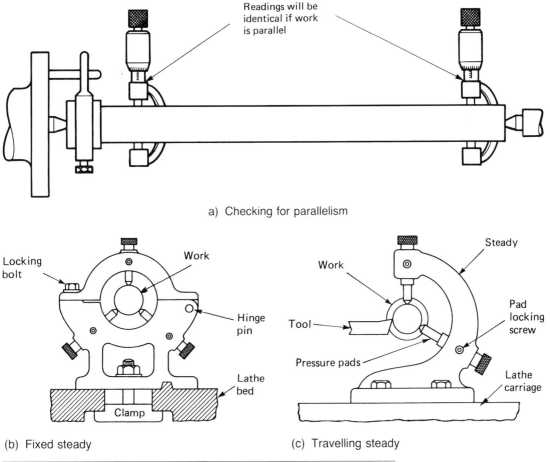

a) Checking for parallelism

(b) Fixed steady

(c) Travelling steady

Figure 6.20 Parallel turning.

Surfacing

A surfacing (facing or perpendicular-turning) operation on a workpiece held in a chuck is shown in Figure 6.21. The saddle is clamped to the bed of the lathe and the tool motion is controlled by the cross-slide. This ensures that the tool moves in a path at right angles to the workpiece axis and produces a plain surface. In-feed of the cutting tool is controlled by micrometer adjustment of the top-slide.

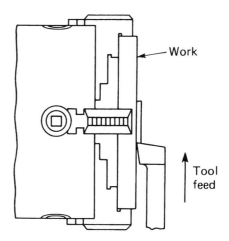

Figure 6.21 Surfacing.

Boring

Figure 6.22 shows how a drilled hole can be opened up using a boring tool. The workpiece is held in a chuck and the tool movement is controlled by the saddle of the lathe. The in-feed of the tool is controlled by micrometer adjustment of the cross-slide. The pilot hole is produced either by a taper shank drill mounted directly into the tailstock barrel (poppet), or by a parallel shank drill

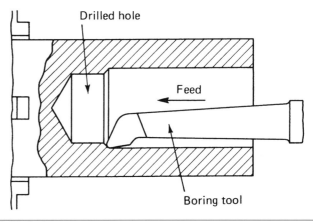

Figure 6.22 Boring.

held in a drill chuck. The taper mandrel of the drill chuck is inserted into the tailstock barrel.

Conical surfaces

Chamfers on the corners of a turned component are short conical surfaces. These are usually produced by using a chamfering tool, as shown in Figure 6.19a. Longer tapers can be produced by use of the top-slide. Use of the top (compound) slide is shown in Figure 6.23. The slide is mounted on a swivel base and it is fitted with a protractor scale. It can be swung round to the required angle and clamped in position. The taper is then cut as shown.

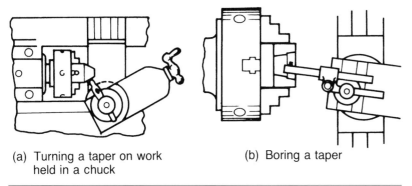

(a) Turning a taper on work held in a chuck

(b) Boring a taper

Figure 6.23 Producing a chamfer.

Miscellaneous turning operations

Reamers are sizing tools. They remove very little metal. Since they follow the existing hole, they cannot correct the positional errors. Hand reamers have a square on the end of their shanks so that they can be rotated by a tap wrench. Machine reamers have a standard Morse taper shank.

Figure 6.24a shows a hole being reamed on a lathe. A machine reamer is being used and it is held in the barrel of the tailstock. Because a drilled hole invariably runs out slightly, the pilot hole should be single-point bored in order to correct the position and geometry of the hole. It is finally sized using the reamer. Only the minimum amount of metal for the surface of the hole to clean up should be left in for the reamer to remove.

Standard, non-standard and large diameter screw threads can be cut in a centre lathe by use of the lead-screw to control the saddle movement. This is a highly skilled operation. However, standard screw threads of limited diameter can be cut using hand threading tools as shown in Figures 6.24b and 6.24c. Note that taps are very fragile and the workpiece should be rotated by hand with the lathe switched off and the gears disengaged.

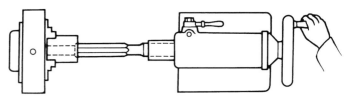

(a) Machine reamer supported in the tailstock

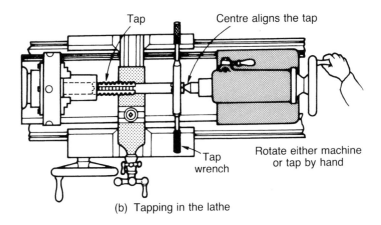

Tap

Centre aligns the tap

Tap wrench

Rotate either machine or tap by hand

(b) Tapping in the lathe

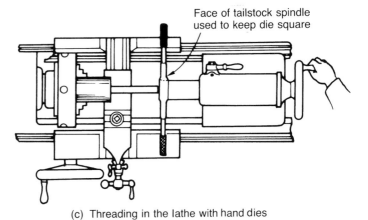

Face of tailstock spindle used to keep die square

(c) Threading in the lathe with hand dies

Figure 6.24 Miscellaneous turning operations.

Learning outcome 6.3

Identify drill bits and their uses

You already know that drilling is a process for producing holes. The holes may be cut from the solid or existing holes may be enlarged. The purpose of the drilling machine is to:

- rotate the drill at a suitable speed for the material being cut and the diameter of the drill
- feed the drill into the workpiece

* support the workpiece being drilled; usually at right angles to the axis of the drill. On some machines the table may be tilted to allow holes to be drilled at a pre-set angle.

Drilling machines

Drilling machines come in a variety of types and sizes. Figure 6.25 shows a hand-held, electrically driven, power drill. It depends upon the skill of the operator to ensure that the drill cuts at right angles to the workpiece. The feed force is also limited to the muscular strength of the user. Figure 6.26 shows a more powerful, floor-mounted machine. The spindle rotates the drill. It can also move up and down in order to feed the drill into the workpiece and withdraw the drill at the end of the cut. Holes are generally produced with twist drills. Figure 6.27 shows a typical straight shank drill and a typical taper shank drill and names their more important features.

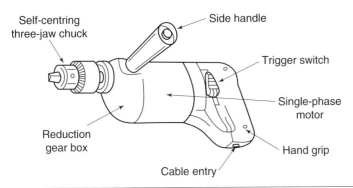

Figure 6.25 An electric power drill.

Large drills have taper shanks and are inserted directly into the spindle of the machine, as shown in Figure 6.28a. They are located and driven by a taper. The tang of the drill is for extraction purposes only. It does not drive the drill. The use of a *drift* to remove the drill is shown in Figure 6.28b.

Small drills have straight (parallel) shanks and are usually held in a self-centring chuck. Such a chuck is shown in Figure 6.28c. The chuck is tightened with the chuck key shown. SAFETY: The chuck key must be removed before starting the machine. The drill chuck has a taper shank which is located in, and driven by, the taper bore of the drilling machine spindle. The cutting edge of a twist drill is wedge-shaped, like all the tools we have considered so far. This is shown in Figure 6.29.

When regrinding a drill it is essential that the point angles are correct. The angles for general purpose drilling are shown in

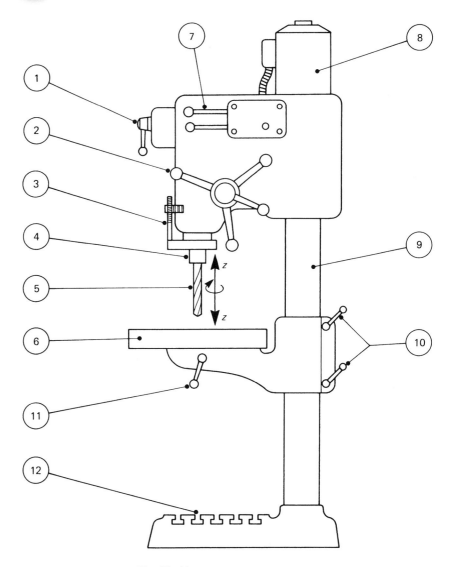

Parts of the Pillar Type Drilling Machine

1	Stop/start switch (electrics).	7	Speed change levers.
2	Hand or automatic feed lever.	8	Motor.
3	Drill depth stop.	9	Pillar.
4	Spindle.	10	Vertical table lock.
5	Drill.	11	Table lock.
6	Table.	12	Base.

Figure 6.26 A pillar drill.

Figure 6.30a. After grinding, the angles and lip lengths must be checked as shown in Figure 6.30b. The point must be symmetrical. The effects of incorrect grinding are shown in Figure 6.30c.

If the lip lengths are unequal, an oversize hole will be drilled when cutting from the solid. If the angles are unequal, then only one

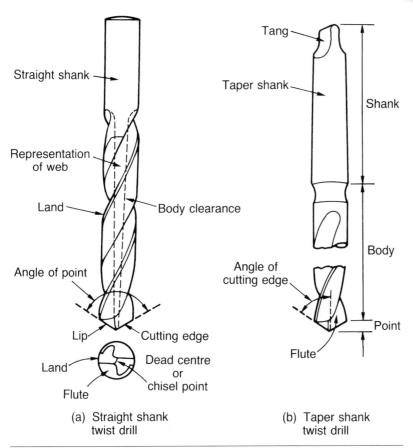

Figure 6.27 Straight shank and taper shank twist drills.

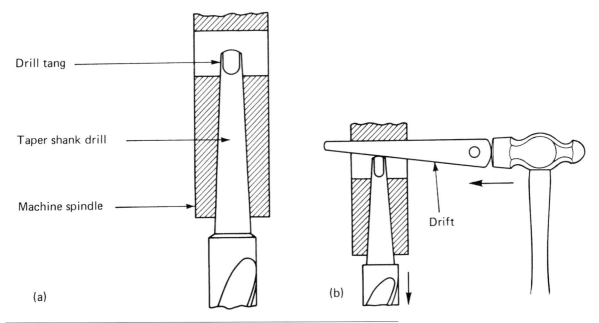

Figure 6.28 Methods of holding a twist drill.

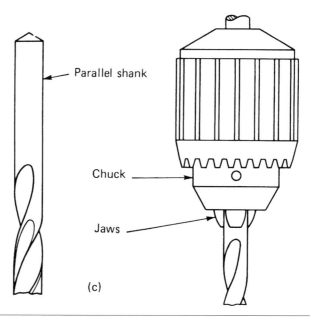

Parallel shank

Chuck

Jaws

(c)

Figure 6.28 (Continued)

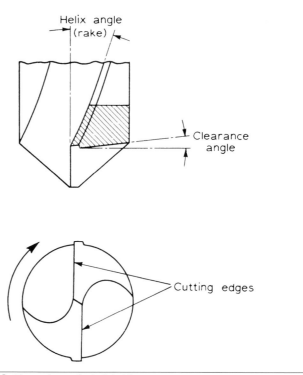

Helix angle
(rake)

Clearance
angle

Cutting edges

Figure 6.29 Cutting edges of a twist drill.

lip will cut and undue wear will result. The unbalanced forces will cause the drill to flex and 'wander'. The axis of the hole will become displaced as drilling proceeds. If both these faults are present at the same time, an inaccurate and ragged hole will result.

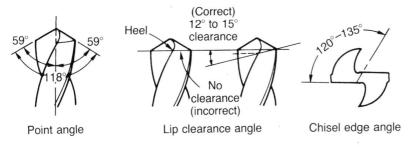

(Correct)
12° to 15°
clearance

Heel

59°   59°

118°

No
clearance
(incorrect)

120°–135°

Point angle   Lip clearance angle   Chisel edge angle

(a) Drill angles for general purpose drilling

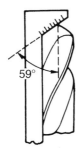

59°

(b) Checking for correct point angle and
equal lip lengths

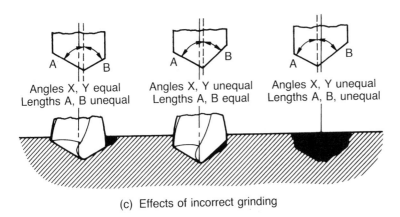

A   B   A   B   A   B

Angles X, Y equal   Angles X, Y unequal   Angles X, Y unequal
Lengths A, B unequal   Lengths A, B equal   Lengths A, B, unequal

(c) Effects of incorrect grinding

Figure 6.30 Point angles for a twist drill.

Work-holding when drilling

It is dangerous to hold work being drilled by hand. There is always
a tendency for the drill to grab the work and spin it round. Also the
rapidly spinning *swarf* can produce some nasty cuts to the back of
your hand. Therefore the work should always be securely fastened
to the machine table. Nevertheless, small holes in relatively large
components are sometime drilled with the work hand-held. In
this case a stop bolted to the machine table should be used to
prevent rotation.

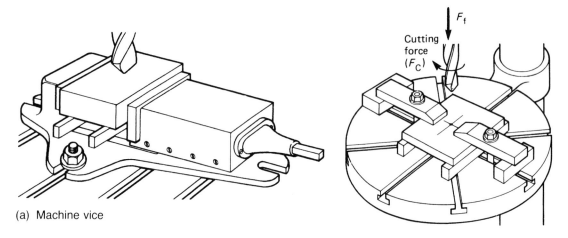

(a) Machine vice

(b) Work supported on parallels and clamped to table

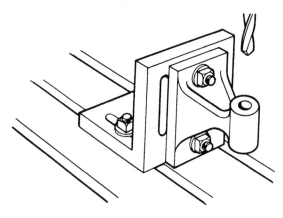

(c) Use of angle plate

Figure 6.31 Work-holding when drilling.

Small work is usually held in a machine vice which, in turn, is securely bolted to the machine table. This is shown in Figure 6.31a. Larger work can be clamped directly to the machine table, as shown in Figure 6.31b. In both these examples the work is supported on parallel blocks. You mount the work in this way so that when the drill 'breaks through' the workpiece it does not damage the vice or the machine table.

Figure 6.31c shows how an angle plate can be used when the hole axis has to be parallel to the datum surface of the work. Figures 6.32a and 6.32b show how cylindrical work is located and supported using vee-blocks. Finally, Figure 6.33 shows some miscellaneous operations that are frequently carried out on drilling machines. These include countersinking, counterboring and spot-facing.

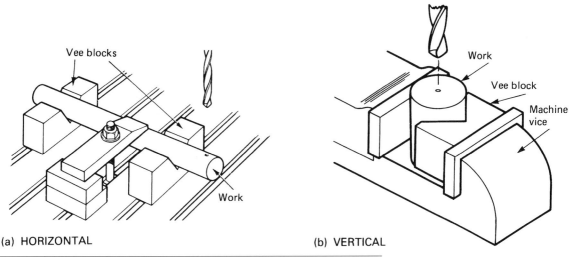

(a) HORIZONTAL   (b) VERTICAL

Figure 6.32 Work-holding cylindrical components.

Countersinking

Figure 6.33a shows a countersink bit being used to countersink a hole to receive the heads of rivets or screws. For this reason the included angle is 90°. Lathe centre drills are unsuitable for this operation as their angle is 60°.

Counterboring

Figure 6.33b shows a piloted counterbore being used to counterbore a hole so that the head of a capscrew or a cheese-head screw can lie below the surface of the work. Unlike a countersink cutter, a counterbore is not self-centring. It has to have a pilot which runs in the previously drilled bolt or screw hole. This keeps the counterbore cutting concentrically with the original hole.

Spot-facing

This is similar to counterboring but the cut is not as deep. It is used to provide a flat surface on a casting or a forging for a nut and washer to 'seat' on. Sometimes, as shown in Figure 6.33c, it is used

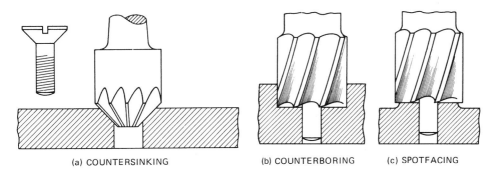

(a) COUNTERSINKING   (b) COUNTERBORING   (c) SPOTFACING

Figure 6.33 Countersinking, counterboring and spot facing.

to machine a *boss* (raised seating) to provide a flat surface for a nut and washer to 'seat' on.

Test your knowledge 6.14

For clarity, no drill guards have been shown in any of the previous examples:

1 Sketch a typical drilling machine spindle/chuck/drill guard assembly.
2 Describe briefly the most common type of accident resulting from the use of an unguarded drilling machine.

Test your knowledge 6.15

Describe one method of locating the work under a drill so that the hole will be cut in the required position.

Test your knowledge 6.16

When drilling explain why:

a) the packing pieces must be parallel and the same size.
b) the work should not be held by hand when drilling large diameter holes.

Key point

Nowadays, manually operated drilling machines (as well as lathes and milling machines) are being increasingly replaced by computer numerically controlled (CNC) machines. These machines can be programmed in order to perform repetitive tasks quickly and accurately.

Learning outcome 6.4

Identify the basic screw thread forms and their uses

The term 'threaded components' refers to nuts, bolts, screws and studs. These come in a wide variety of sizes and types of screw thread. Figure 6.34 shows some typical threaded component fastenings and it also shows how they are used. When selecting a threaded component for any particular purpose, you should ask yourself the following questions.

• Is the component strong enough for the application?
• Is the material from which the fastening is made corrosion-resistant under actual service conditions and is it compatible with the metals being joined?
• Is the screw thread chosen suitable for the job? Coarse threads are stronger than fine threads, particularly in soft metals such as aluminium. Fine threads are less likely to work loose.

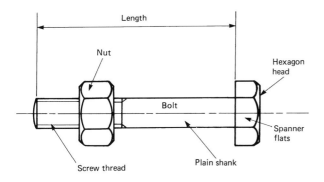

(a) HEXAGON HEAD BOLT AND NUT

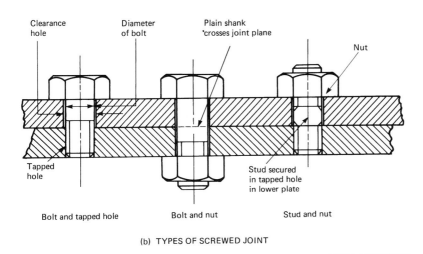

(b) TYPES OF SCREWED JOINT

Figure 6.34 Some typical threaded fastenings.

There are a large variety of heads for threaded fastenings, and the selection is usually a compromise between strength, appearance and ease of tightening. The hexagon head is usually selected for general engineering applications. The more expensive cap-head bolt is widely used in the manufacture of machine tools, jigs and fixtures, and other highly stressed applications. These fastenings are forged from high-tensile alloy steels, thread rolled and heat treated. By recessing the cap-head, a flush surface is provided for safety and easy cleaning. Figure 6.35 shows some alternative screw heads.

Thread sizes

Metric bolt sizes are specified in millimetres (mm) and include the diameter, pitch and length of a bolt. Metric bolt specification begins

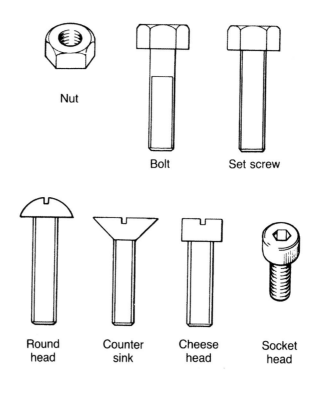

Nut

Bolt

Set screw

Round head

Counter sink

Cheese head

Socket head

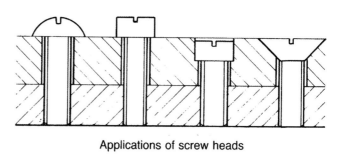

Applications of screw heads

Figure 6.35 Various types of nut, bolt and screw.

with the letter 'M' and the pitch of a screw thread is simply the distance measured between adjacent peaks or troughs.

Diameter

The first number in a metric bolt specification gives the nominal outer diameter of the bolt. For example, a bolt specified as 'M8-1.25×30' fits an 8mm diameter hole. It is important to be aware that metric bolts are usually very slightly smaller than their specified diameter. This difference in specified and actual diameter allows for a small clearance between the screw and the hole through which it passes and allows some flexibility for parts that might be misaligned. The strength of a bolt will increase with its diameter but the material used to manufacture the bolt also has an impact on its strength. For example, hardened steel bolts are stronger than stainless steel bolts of the same diameter.

Pitch

The pitch of a bolt refers to the distance between two adjacent threads. Pitch is specified in mm and an 'M8-1.25×30' bolt will have a pitch of 1.25mm. If you need to convert pitch to threads-per-inch (TPI) you can simply divide 25.4 by the metric pitch in mm. For example, a bolt with a 1.25mm pitch will have 25.4/1.25 = 20.3 threads-per-inch.

Pitch is often also described as either 'fine' or 'standard/coarse'. Metric fine bolts generally have a pitch of 1.5mm or smaller. The most common larger-size metric bolts have a coarse thread. Some common diameter and pitch combinations are shown in Table 6.4. Fine pitch threads are often used when a bolt is screwed into a soft metal block, such as an alloy casting. It is very important to be aware that the thread in the block may be much weaker than the bolt and a lower torque is almost always required to avoid stripping the threads.

Key point

Fine pitch nuts and bolts are normally only used in special circumstances and the manufacturer's documentation should be consulted in order to ascertain the maximum value of torque that should be applied.

Table 6.4 Common metric thread diameter and pitch combinations.

Metric size	Standard pitch (mm)	Fine pitch (mm)
M6	1.0	0.75
M8	1.25	1.0
M10	1.5	1.0
M12	1.75	1.5
M16	2.0	1.5
M20	2.5	1.5
M24	3.0	2.0

Key point

Fine pitch threads are often used when a hard metal bolt is screwed into a soft metal block, such as an alloy casting. When tightening, care must be taken to limit the applied torque in order to avoid stripping the thread made in the softer metal.

Length

The final number in the specification of a metric bolt gives its length. So, for example, an 'M8-1.25×30' bolt will have a length of 30mm. Most bolts, including cheese, hex, pan, socket, button and low socket head types (see Figure 6.35) are measured from under the head to the tip of the bolt. However, the length of oval and flat head bolts includes their head height.

Strength

Metric bolts (and nuts) may carry markings which indicate their *ultimate tensile strength*. This is the amount of tensile stress that a part can withstand before it fractures. Strength markings are usually shown on the head of the bolt and take the form of two numbers separated by a decimal point. Note that nuts and bolts may also carry a two- or three-letter marking to indicate the manufacturer.

The first number on the strength marking gives the breaking strength of the material used to manufacture the bolt expressed in tens of kilogram-force (kgf) per square millimeter (mm) of the bolt. As an example, a bolt marked '10.8' can be expected to sustain a load of up to 100 kgf per square mm before it breaks. However, before this occurs (at less load) the bolt may begin to elongate before it finally fails. The second figure gives the proportion of the load at which this begins to occur. So, with a bolt marked '10.8', the load at which stretching becomes a problem will be 0.8 or 80% of its specified breaking load.

In manufacturers' data a *proof strength* rating may be quoted as well as ultimate tensile strength. This is the maximum amount of tensile stress that a part can withstand before it begins to exhibit a permanent deformation. In other words, the part will no longer return to its original shape after the tensile force has been removed. It is also worth noting that manufacturers' data often expresses strength in megapascals (MPa) rather than kilogram-force (kgf) per square millimetre. Because $1\,\text{kgf/mm}^2 = 9.81\,\text{MPa}$ you can easily convert from kgf/mm^2 to MPa by simply multiplying by 9.81.

Strength grade markings on nuts can often be difficult to see. They may either be stamped on the top of the nut or on one of the flat sides. Clock face grade marks are sometimes also used. When selecting a nut the general rule is that the grade of the nut should always match the grade of the bolt or be one grade higher.

> **Key point**
>
> You can convert from kgf/mm^2 to MPa by multiplying by 9.81. To convert from MPa to kgf/mm^2 you can multiply by 0.102.

> **Key point**
>
> The proof strength of a component is the maximum amount of tensile stress that a part can withstand before it begins to exhibit a permanent deformation. The ultimate tensile strength of a part is the maximum amount of tensile strength that can be applied to the component before it breaks.

Test your knowledge 6.17

Explain what is meant by the 'proof strength' of a component. How does this differ from the ultimate tensile strength?

Test your knowledge 6.18

A bolt head is marked '12.9'. Explain what this means and determine the ultimate tensile strength of the bolt expressed in MPa.

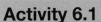

Activity 6.1

Bolts have maximum recommended torque values that must not be exceeded when tightening. The maximum torque rating is usually equivalent to about 85% of the proof load (the point at which the bolt will begin to permanently elongate). Use the Internet to obtain the maximum recommended torque for a) an M10 bolt with an '8.8' strength rating and b) an M10 bolt with a '10.9' strength rating. You will find more information on the turning effect of a force on page 296.

Learning outcome 6.5

Identify the basic methods of work assembly

Engineered products usually comprise a number of components that must be assembled or joined together in some particular way. The purpose of assembly is to put together a number of individual components to build up a whole device, structure or system. To achieve this aim, attention must be paid to the following key factors.

Sequence of assembly

This must be planned so that as each component is added, its position in the assembly and the position of its fastenings are accessible. Also the sequence of assembly must be planned so that the installation of one component does not prevent access for fitting the next component or some later component.

Technique of joining

These must be selected to suit the components being joined, the materials from which they are made and what they do in service. If the joining technique involves heating, then care must be taken that adjacent components are not heat sensitive or flammable.

Position of joints

Joints must not only be accessible for initial assembly, they must also be accessible for maintenance. You don't want to dismantle half a machine to make a small adjustment, or replace a part that wears out regularly!

Interrelationship and identification of parts

Identification of parts and their position in an assembly can usually be determined from assembly drawings or exploded view drawings. Interrelationship markings are often included on components. For example, the various members and joints of structural steelwork are given number and letter codes to help identification on site. Printed circuit boards often have a silk screen printed outline of the various components as well as their part numbers.

Tolerances

The assembly technique must take into account the accuracy and finish of the components being assembled. Much greater care has to be taken when assembling a precision machine tool or an artificial satellite, than when assembling structural steel work.

Protection of parts

Components awaiting assembly require protection against accidental damage and corrosion. In the case of structural steelwork this may merely consist of painting with red oxide primer and careful stacking. Precision components will require treating with an anti-corrosion lanolin-based compound that can be easily removed at the time of assembly. Bores must be sealed with plastic plugs and screw threads with plastic caps. Precision ground surfaces must also be protected from damage. Heavy components must be provided with eye-bolts for lifting. Vulnerable sub-assemblies such as aircraft engines must be supported in suitable cradles.

Joining

The joints used in engineering assemblies may be divided into several categories including permanent joints, temporary joints and flexible joints.

Permanent joints

These are joints in which one or more of the components and/or the joining medium has to be destroyed or damaged in order to dismantle the assembly – for example, a riveted joint.

Temporary joints

Temporary joints are used to assemble parts that need to be dismantled and re-assembled when required. This can be important when maintenance is necessary. These are joints that can be dismantled without damage to the components. It should be possible to re-assemble the components using the original or new fastenings – for example, a bolted joint.

Flexible joints

These are joints in which one component can be moved relative to another component in an assembly in a controlled manner – for example, the use of a hinge.

Test your knowledge 6.19

Explain why it is necessary to protect a joint made between the structural components used on an oil rig. Describe one method of protection.

Assembly using threaded fastenings

Figure 6.36 shows the correct way to use some typical threaded fastenings. Bolted and screwed fastenings must always pull down onto prepared seatings that are flat and at right angles to the axis of the fastening. This prevents the bolt or screw being bent as it

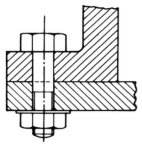

(a) Section through a bolted joint (plain shank extends beyond joint face)

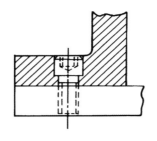

(b) Cap head socket screw (head recessed into counterbore to provide flush surface)

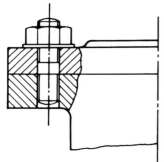

(c) Stud & nut fixing for inspection cover (used where joint has to be regularly dismantled)

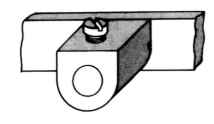

(d) Cheese head brass screw (for clamping electrical conductor into terminal)

Figure 6.36 Screwed fastenings.

is tightened up. To protect the seating, a soft washer is placed between the seating and the nut. Taper washers are used when erecting steel girders to prevent the draught angle of the flanges from bending the bolt.

Locking devices are used to prevent threaded fastenings from slackening off due to vibration. Locking devices may be frictional or positive. A selection of plain washers, taper washers and locking devices is shown in Figure 6.37.

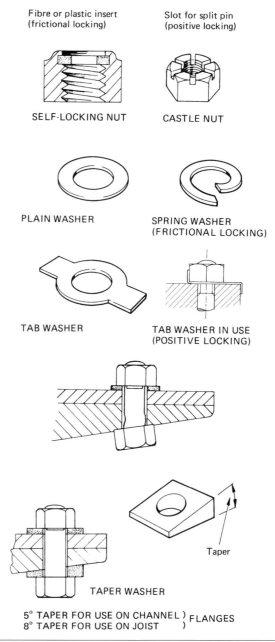

Figure 6.37 Various nuts and washers.

Assembly using riveted fastenings

Figure 6.38 shows some typical riveted joints. Riveted joints are very strong providing they are correctly designed and assembled. The joint must be designed so that the rivet is in shear and not in tension. Consider the head of the rivet as only being strong enough to keep the rivet in place. You must consider a number of factors when selecting a rivet and making a riveted joint. These include the material used for the rivet as well as the shape of its head. The material from which the rivet is made must not react with the components being joined as this will cause corrosion and weakening. Also the rivet must be strong enough to resist the loads imposed upon it. The rivet head chosen is always a compromise between strength and appearance. In the case of aircraft components, wind resistance must also be taken into account. Figure 6.39 shows some typical rivet heads and rivet types.

Making a riveted joint

To make a satisfactory riveted joint the following points must be observed:

- *Hole clearance.* If the clearance is too small there will be difficulty in inserting the rivet and drawing up the joint. If the hole is too large, the rivet will buckle and a weak joint will result.
- *Rivet length.* If the rivet is too long, the rivet bends over during heading. If the rivet is too short, the head cannot be properly formed. In either case a weak joint will result. Figure 6.40a shows the correct proportions for a riveted joint and Figure 6.40b shows some typical riveting faults.

The correct procedure for heading (closing) a rivet is shown in Figure 6.41a. The drawing-up tool ensures that the components to be drawn are brought into close contact and that the head of the rivet is drawn up tightly against the lower component. The hammer blows with the flat face of the hammer swells the rivet so that it is a tight fit in the hole and starts to form the head. The ball pein of the hammer head is then used to rough form the rivet head. The head is finally finished and made smooth by using a rivet snap. Where large rivets and large quantities of rivets are to be closed a portable pneumatic riveting tool is used.

'Pop' riveting is often used for joining thin sheet metal components, particularly when building up box sections. When building up box sections it is not possible to get inside the closed box to use a hold-up dolly – for example, when riveting the skin to an aircraft wing. Figure 6.41b shows the principle of 'pop' riveting.

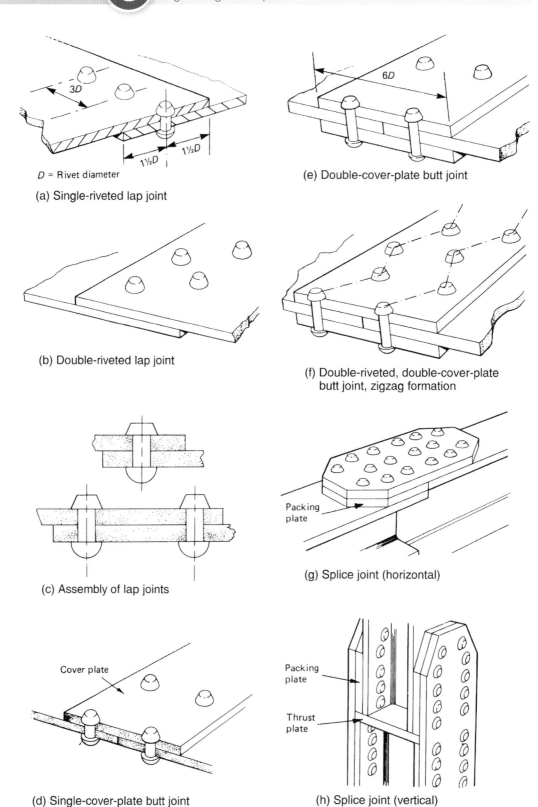

D = Rivet diameter

(a) Single-riveted lap joint

(b) Double-riveted lap joint

(c) Assembly of lap joints

(d) Single-cover-plate butt joint

(e) Double-cover-plate butt joint

(f) Double-riveted, double-cover-plate butt joint, zigzag formation

(g) Splice joint (horizontal)

(h) Splice joint (vertical)

Figure 6.38 Typical riveted joints.

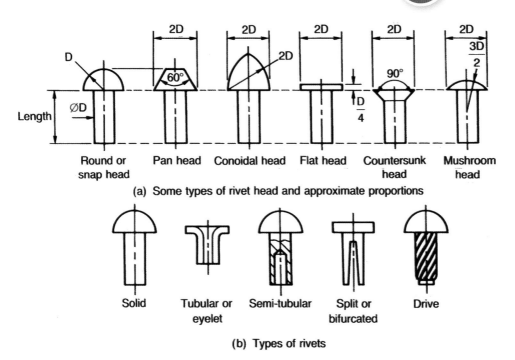

Length

ØD

Round or snap head | **Pan head** | **Conoidal head** | **Flat head** | **Countersunk head** | **Mushroom head**

(a) Some types of rivet head and approximate proportions

Solid | **Tubular or eyelet** | **Semi-tubular** | **Split or bifurcated** | **Drive**

(b) Types of rivets

Figure 6.39 Typical rivet heads and types.

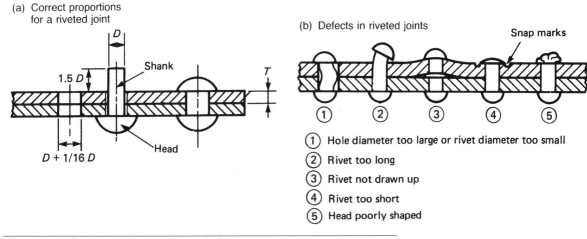

(a) Correct proportions for a riveted joint

Shank

1.5 D

D + 1/16 D

Head

(b) Defects in riveted joints

Snap marks

① Hole diameter too large or rivet diameter too small
② Rivet too long
③ Rivet not drawn up
④ Rivet too short
⑤ Head poorly shaped

Figure 6.40 Correct and incorrect riveted joints.

Test your knowledge 6.20

List two important factors that need to be taken into account when preparing a riveted joint between two flat plates.

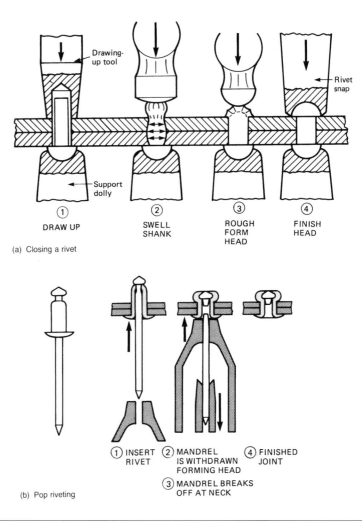

Figure 6.41 Correct procedure for riveting.

Activity 6.2

Obtain a portable electric fan heater. Carefully dismantle the heater (make sure that it is disconnected from AC supply when you do this) and examine the types of fastening that are used. Write a brief report describing the fastenings and include sketches or photographs.

Welded joints

Welding has largely taken over from riveting for many purposes such as ship and bridge building and for structural steelwork. Welded joints are continuous and, therefore, transmit the stresses across the joint uniformly. In riveted joints the stresses are concentrated at each rivet. Also the rivet holes reduce the cross-sectional areas of the members being joined and weaken them. However, welding is a more skilled assembly technique and the equipment required is more costly. The components being joined are melted at their edges and additional filler metal is melted into the joint. The filler metal is of similar composition to that of the components being joined. Figure 6.42a shows the principle of fusion welding.

Oxy-acetylene welding

High temperatures are involved to melt the metal of the components being joined. These can be achieved by using the flame of an oxy-acetylene blowpipe, as shown in Figure 6.42b, or an electric arc, as shown in Figure 6.42c. When oxy-acetylene welding (gas welding), a separate filler rod is used. When arc welding, the electrode is also the filler rod and is melted as welding proceeds.

No flux is required when oxy-acetylene welding as the molten metal is protected from atmospheric oxygen by the burnt gases (products of combustion). When arc welding, a flux is required. This is in the form of a coating surrounding the electrode. This flux coating is not only deposited on the weld to protect it, it also stabilizes the arc and makes the process easier. The hot flux gives off fumes and adequate ventilation is required.

Protective clothing must always be worn when welding and goggles or a face mask (visor) appropriate for the process must be used. These have optical filters that protect the user's eyes from the harmful radiations produced during welding. The optical filters must match the process.

The compressed gases used in welding are very dangerous and welding equipment must only be used by skilled persons or under close supervision. Acetylene gas bottles must only be stored and used in an upright position.

The heated area of the weld is called the weld zone. Because of the high temperatures involved, the heat-affected area can spread back into the parent metal of the component for some distance from the actual weld zone. This can alter the structure and properties of the material so as to weaken it and make it more brittle. If the joint fails in service, failure usually occurs at the side of the weld in this heat-affected zone. The joint itself rarely fails.

Test your knowledge 6.21

When joining metal parts together, state **three** advantages and **three** disadvantages of welding compared with riveting.

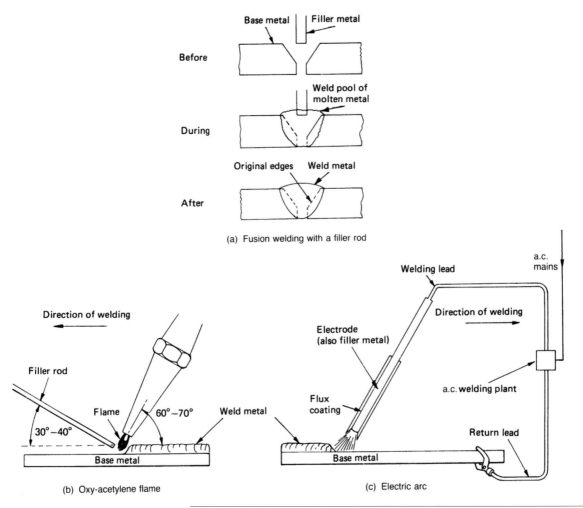

(a) Fusion welding with a filler rod

(b) Oxy-acetylene flame

(c) Electric arc

Figure 6.42 Fusion welding.

Soldering

Like welding, soldering is also a thermal jointing process. Unlike welding, the parent metal is not melted and the filler metal is an alloy of tin and lead that melts at relatively low temperatures. Soft soldering is mainly used for making mechanical joints in copper and brass components (plumbing). It is also used to make permanent electrical connections. Low carbon steels can also be soldered providing the metal is first cleaned and then *tinned* using a suitable flux. The tin in the solder reacts chemically with the surface of the component to form a bond.

Figure 6.43 shows how to make a soft soldered joint. The surfaces to be joined are first degreased and physically cleaned to remove any dust and dirt. Fine abrasive cloth or steel wool can be used. A *flux* is used to render the joint surfaces chemically clean and to make the solder spread evenly through the joint.

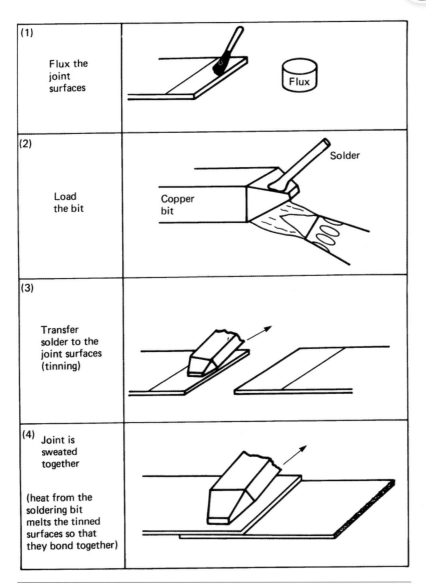

(1) Flux the joint surfaces	
(2) Load the bit	Copper bit, Solder
(3) Transfer solder to the joint surfaces (tinning)	
(4) Joint is sweated together (heat from the soldering bit melts the tinned surfaces so that they bond together)	

Figure 6.43 Procedure for making a soft soldered joint.

- The copper *bit* of the soldering iron is then heated. For small components and fine electrical work an electrically heated iron can be used. For joints requiring a soldering iron with a larger bit, a gas heated soldering stove can be used to heat the bit.
- The heated bit is then cleaned, fluxed and coated with solder. This is called *tinning* the bit.
- The heated and tinned bit is drawn slowly along the fluxed surfaces of the components to be joined. This transfers solder to the surfaces of the components. Additional solder can be added if required. The work should be supported on wood to prevent heat loss. The solder does not just 'stick' to the surface of the metal being tinned. The solder reacts chemically with the surface

to form an amalgam that penetrates into the surface of the metal. This forms a permanent bond.

- Finally the surfaces are overlapped and 'sweated' together. That is, the soldering iron is reheated and drawn along the joint as shown. Downward pressure is applied at the same time. The solder in the joint melts. When it solidifies it forms a bond between the two components.

Figure 6.44 shows how a copper pipe is sweated to a fitting. The pipe and the fitting are cleaned, fluxed and assembled. The joint is heated with a propane gas torch and solder is added. This is usually a resin-flux cored solder. The solder is drawn into the close-fitting joint by capillary action.

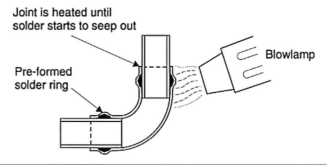

Figure 6.44 'Sweating' a copper pipe.

Activity 6.3

Consult manufacturers' or suppliers' data and draw up a table showing the composition of several soft solders and their typical applications.

Hard soldering

Hard soldering uses a solder whose main alloying elements are copper and silver. Hard soldering alloys have a much higher melting temperature range than soft solders. The melting range for a typical soft solder is 183°C to 212°C. The melting range for a typical hard solder is 620°C to 680°C – this makes hard soldering unsuitable for use with electronic components.

Hard soldering produces joints that are stronger and more ductile. The melting range for hard solders is very much lower than the melting point of copper and steel, but it is only just below the melting point of brass. Therefore great care is required when hard soldering brass to copper. Because the hard solder contains silver it is often referred to as 'silver solder'. A special flux is required based on *borax* (sodium borate).

Since high temperatures are involved a soldering iron cannot be used. Instead, heating is usually applied by means of a blow pipe. Figure 6.45 shows you how to make a typical hard soldered joint. Again, cleanliness and careful surface preparation is essential for a successful joint. The joint must be close-fitting and free from voids. The silver solder is drawn into the joint by capillary action.

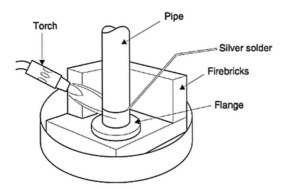

- The work is up to heat when the silver solder melts on contact with the work with the flame momentarily withdrawn.
- Add solder as required until joint is complete.

Figure 6.45 Procedure for making a hard soldered joint.

Test your knowledge 6.22

List the advantages and limitations of soft soldering compared with hard soldering.

Brazing

Even stronger joints can be made using a brass alloy instead of a silver-copper alloy. This is called *brazing*. The temperatures involved are higher than those for silver soldering. Therefore, brass cannot be brazed. The process of brazing is widely used for joining steel tubes and malleable cast iron fittings.

Since brazing does not melt the parent metal, less heat is required, but the resulting joint is not as strong as a welded joint. The brazing alloy is often a different colour from the parent metal, so the joint will stand out and be visible. This can be a problem where appearance is important.

This filler is a *brazing alloy* which often contains brass, silver, or other metals. Note that the filler material is not the same as the parent metal.

Adhesive bonding

The advantages of adhesive bonding can be summarized as follows.

Key point

In welding, the parent metal in the parts to be joined is melted along with the filler metal (if any). The filler metal is the same material as the parent metal. In brazing, the parts to be joined are heated but they do not melt. Only the filler is melted.

Key point

The basic difference between soldering and brazing is the temperature necessary to melt the filler metal. That temperature is usually defined as 450°C. So, if the filler material melts below 450°C, the process is soldering.

- The temperature rise from the curing of the adhesive is negligible compared with that of welding. Therefore the properties of the materials being joined are unaffected.
- Similar and dissimilar materials can be joined.
- Adhesives are electrical insulators. Therefore they reduce or prevent electrolytic corrosion when dissimilar metals are joined together.
- Joints are sealed and fluid tight.
- Stresses are transmitted across the joint uniformly.
- Depending upon the type of adhesive used, some bonded joints tend to damp out vibrations.

Bonded joints have to be specially designed to exploit the properties of the adhesive being used. You cannot just substitute an adhesive in a joint designed for welding, brazing or soldering. Figure 6.46 shows some typical bonded joint designs that provide a large contact area. A correctly designed bonded joint is very strong. Major structural members in modern high-performance airliners and military aircraft are adhesive bonded. The strength of adhesive bonded joints like those shown in Figure 6.46 depends on two important factors, adhesion and cohesion.

Adhesion

This is the ability of the adhesive to 'stick' to the materials being joined (the *adherends*). This can result from physical keying or interlocking, as shown in Figure 6.47a. Alternatively specific bonding can take place. Here, the adhesive reacts chemically with the surface of the adherends, as shown in Figure 6.47b. Bonding occurs through intermolecular attraction.

Cohesion

This is the internal strength of the adhesive. It is the ability of the adhesive to withstand forces within itself. Figure 6.47c shows the failure of a joint made from an adhesive that is strong in adhesion but weak in cohesion. Figure 6.47d shows the failure of a joint that is strong in cohesion but weak in adhesion.

As well as the design of the joint, the following factors affect the strength of a bonded joint:

- The joint must be physically clean and free from dust, dirt, moisture, oil and grease.
- The joint must be chemically clean. The materials being joined must be free from scale or oxide films.
- The environment in which bonding takes place must have the correct humidity and be at the correct temperature.

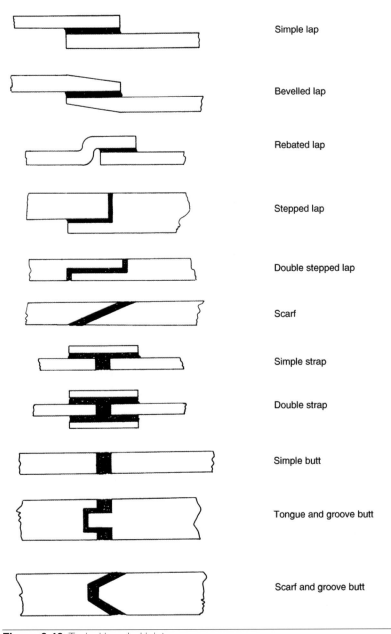

Figure 6.46 Typical bonded joints.

Bonded joints may fail in four ways. These are shown in Figure 6.48. Bonded joints are least likely to fail in tension and shear. They are most likely to fail in cleavage and peel. The most efficient way to apply adhesives is by an adhesive gun. This enables the correct amount of adhesive to be applied to the correct place without wastage or mess. It also prevents the evaporation of highly flammable and toxic solvents while the adhesive is waiting to be used.

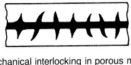

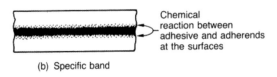

(a) Mechanical interlocking in porous materials

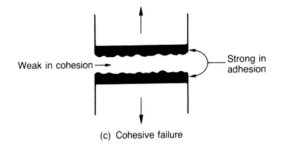

Chemical reaction between adhesive and adherends at the surfaces

(b) Specific band

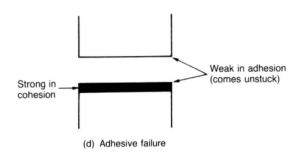

Weak in cohesion →

Strong in adhesion

(c) Cohesive failure

Strong in cohesion

Weak in adhesion (comes unstuck)

(d) Adhesive failure

Figure 6.47 Adhesion.

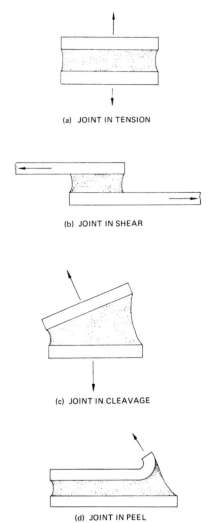

(a) JOINT IN TENSION

(b) JOINT IN SHEAR

(c) JOINT IN CLEAVAGE

(d) JOINT IN PEEL

Figure 6.48 Ways in which bonded joints can fail.

Test your knowledge 6.23

Explain the meaning of the terms 'adhesion' and 'cohesion' when referring to adhesive joining techniques.

Test your knowledge 6.24

State the precautions necessary during the design stage and during the manufacture of an adhesive bonded joint.

Joining electrical and electronic components

When joining electrical and electronic components the joint needs to be reliable, stable and able to conduct electric current easily. It is also important to avoid overheating small electronic components during the soldering process. Overheating can soften thermoplastic insulation and may completely destroy solid state devices such as diodes, transistors and integrated circuits. Soldering can be performed manually or using automated manufacturing techniques in which machines place components on printed circuit boards prior to soldering using a solder bath (see later).

A high tin content, low melting temperature solder with a resin flux core should be used. This is a passive flux. It only protects the joint. It contains no active, corrosive chemicals to clean the joint. Therefore the joint must be kept clean while soldering. Even the natural grease from your fingers is sufficient to cause a high resistance 'dry' joint.

Figure 6.49a shows how a soldered connection is made to a solder tag. Note how the lead from the resistor is secured around the tag before soldering. This gives mechanical strength to the connection. Soldering provides the electrical continuity.

Figure 6.49b shows a prototype electronic circuit assembled on a matrix board. The board is made from laminated plastic and is pierced with a matrix of equally spaced holes. Pin tags are fastened into these holes in convenient places and the components are soldered to these pin tags.

Figure 6.49c shows the same circuit built up on a strip board. This is a laminated plastic board with copper tracks on the underside. The wire tails from the components pass through the holes in the board and are soldered to the tracks on the underside. The copper tracks are cut wherever a break in the circuit is required.

Figure 6.49d shows the underside of a printed circuit board (PCB). This is built up as shown in Figure 6.49c, except that the tracks do not need to be cut since they are already printed for the circuit in question with copper tracks and 'land' areas under the board.

Large volume assembly of printed circuit boards involves the use of pick-and-place robots to install the components. The assembled boards are then carried over a flow soldering tank on a conveyor. A roller rotates in the molten solder creating a 'hump' in the surface of the solder. As the assembled and fluxed board passes over this 'hump' of molten solder the components tags are soldered into place.

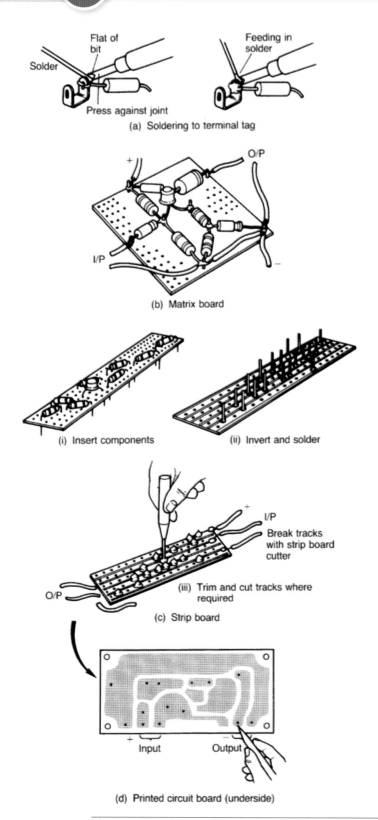

Figure 6.49 Various methods of electronic circuit assembly.

Wire wrapping

Wire wrapping is widely used in telecommunications where large numbers of fine conductors have to be terminated quickly and in close proximity to each other. Soldering would be inconvenient and the heat could damage the insulation of adjoining conductors. Also soldered joints would be difficult to disconnect. A special tool is used that automatically strips the insulation from the wire and binds the wire tightly around the terminal pins. The terminal pins are square in section with sharp corners. The corners cut into the conductor and prevents it from unwinding. The number of turns round the terminal is specified by the supervising engineer.

Crimped joints

For power circuits, particularly in the automotive industry, cable lugs and plugs are crimped onto the cables. The sleeve of the lug or the plug is slipped over the cable and then indented by a small pneumatic or hydraulic press. This is quicker than soldering and, as no heat is involved, there is no danger of damaging the insulation. Portable equipment is also available for making crimped joints on site. Hand-operated equipment can be used to fasten lugs to small cables by crimping, as shown in Figure 6.50.

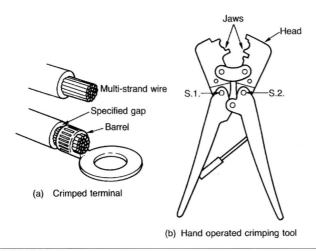

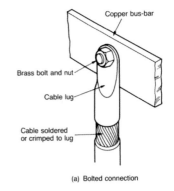

(a) Bolted connection

Figure 6.50 Crimping.

Clamped connections

Finally, we come to clamped connections using screwed fastenings. You will have seen many of these in domestic plugs, switches and lamp-holders. For heavier power installations, cable lugs are bolted to solid copper bus-bars using brass or bronze bolts, as shown in Figure 6.51.

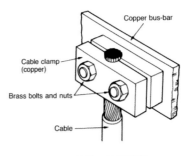

(b) Clamped connection

Figure 6.51 Bolted and clamped connections.

Test your knowledge 6.25

Describe **three** different methods of connecting electrical/ electronic components.

Activity 6.4

A manufacturer of electronic kits has asked you to produce a single page instruction sheet on 'How to solder your kit'. Produce a word-processed instruction sheet and illustrate it using appropriate sketches and drawings. Include a section headed 'Safety'.

Learning outcome 6.6

Identify basic fault-finding techniques to simple problems

Fault-finding is a disciplined and logical process in which 'trial fixing' should never be contemplated. The generalized process of fault-finding is illustrated in the flowchart of Figure 6.52.

First, you need to verify that the equipment or component really is faulty and that you haven't overlooked something obvious (such as a missing part, defective battery or disconnected power source). This may sound rather obvious but in some cases a fault may simply be attributable to maladjustment or misconnection. Furthermore, where several component parts are connected together, it may not be easy to pinpoint the single item that is faulty.

The second stage is that of gathering all relevant information. This process involves asking questions such as:

• In what circumstances did the equipment or component fail?
• Has the equipment or component operated correctly before and exactly what has changed?
• Is there a service or maintenance record for the equipment or component? If so, when was it last serviced and what was revealed?
• Has the deterioration in performance been sudden or progressive?
• What are the symptoms of the fault and do they change with time?

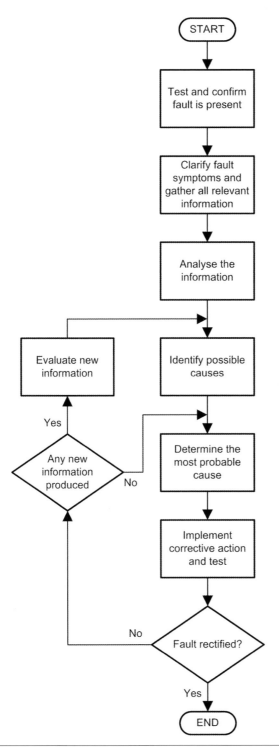

Figure 6.52 General process of fault-finding.

- Is the failure catastrophic (such that the equipment or component does not work at all) or is the fault intermittent?
- Is it possible to predict the circumstances in which the equipment or component will fail and what are they?
- Is it possible to substitute any of the component parts so that known working parts can be used to replace those that are suspect?

The answers to these questions are crucial and, once the information has been analysed, the next stage involves separating the 'effects' from the 'cause'. Here you should list each of the possible causes. Once this has been done, you should be able to identify and focus upon the most probable cause.

Corrective action (such as component removal and replacement, adjustment or alignment) can then be applied before further functional checks are carried out. It should then be possible to determine whether or not the fault has been correctly identified. Note, however, that the failure of one component can often result in the malfunction or complete failure of another. As an example, a short-circuit capacitor in a power supply will often cause the supply fuse to blow. When the fuse is replaced and the supply is reconnected the fuse will once again blow because the capacitor is still faulty. It is therefore important to consider what other problems may be present when a fault is located.

Various information sources can be invaluable when carrying out fault-finding. These include detailed specifications and drawings of the equipment and component, operating manuals and maintenance/service manuals. Some manufacturers also provide detailed fault-finding information in the form of tables and charts, like that shown in Figure 6.53.

Finally, and *before* attempting to carry out *any* fault-finding it is essential to observe relevant safety precautions. In particular you need to be fully aware of the potential hazards associated with equipment that you are working on. Be particularly careful with equipment that uses moving parts, fluids, high voltages etc. In particular, when fault-finding on electronic equipment it is essential to switch off the mains supply *and* disconnect it from the supply before attempting to dismantle the equipment. If, at some later point, you do have to operate equipment with the supply connected you should check that the equipment is properly earthed, avoid direct contact with incoming mains wiring, only use insulated test prods and adjusting tools, and select appropriate meter ranges *before* attempting to take any measurements. If in any doubt about what you are doing, switch off at the mains, disconnect the mains connector and *think!*

Key point

Before attempting to carry out any fault-finding it is essential to observe relevant safety precautions and be fully aware of the potential hazards associated with equipment that you are working on.

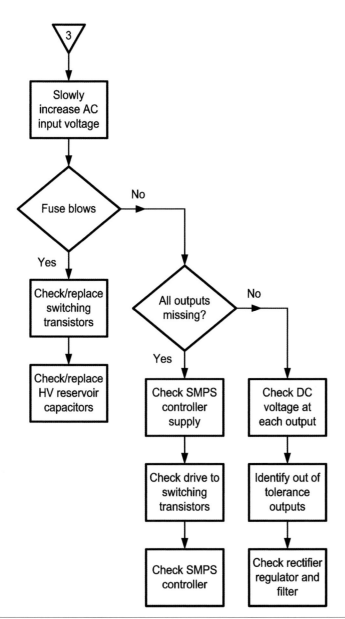

Figure 6.53 Example of a fault-finding chart for an electronic power supply.

Test your knowledge 6.26

Using the fault-finding chart shown in Figure 6.53, what should you do next if you find some of the outputs are working normally and the fuse has not blown?

Review questions

1. Explain what is meant by the 'rake angle' and 'clearance angle' of a cutting tool. Illustrate your answer with a sketch.

2. Explain what is meant by the 'set' of a hacksaw blade. Illustrate your answer with a sketch.

3. Describe the essential differences between soldering and brazing.

4. Explain why it is essential to avoid breaking a tap when tapping a hole. With the aid of a sketch, show how a tap can be held in a centre lathe to reduce the risk of breakage.

5. Explain, with the aid of sketches, the difference between countersinking, counterboring and spot facing.

6. List the safety precautions that must be taken when using oxy-acetylene welding equipment.

7. A bolt is described as 'M5 × 0.8'. Explain what this means.

8. Explain why it is necessary to limit the torque applied when tightening a nut.

9. With the aid of sketches describe **three** faults that can occur when making riveted joints.

10. With the aid of sketches, list **four** ways in which an adhesive bonded joint may fail. Which two of these are most likely to occur?

11. Explain what you would do (and why you would do it) before attempting to fault-find on a piece of equipment that is electrically powered.

Chapter checklist

Learning outcome	Page number
6.1 Identify work- and tool-holding methods and applications.	124
6.2 Identify cutting tool types and their uses.	125
6.3 Identify drill bits and their uses.	146
6.4 Identify the basic screw thread forms and their uses.	154
6.5 Identify the basic methods of work assembly.	159
6.6 Identify basic fault-finding techniques to simple problems.	178

CHAPTER 7

Measurements and marking out

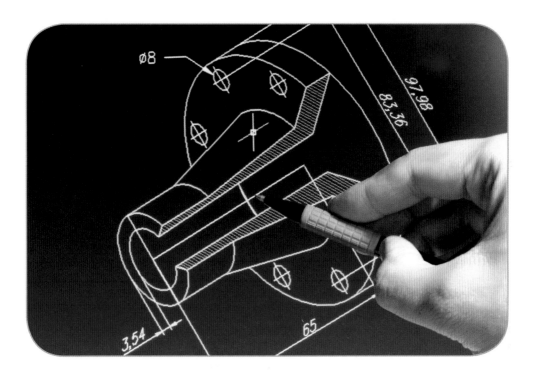

When you have completed this chapter you should understand measurement and marking-out techniques. You should be able to:

7.1 Identify the sources of information used in engineering to support marking-out activities.

7.2 Identify the range of measuring and marking-out equipment available.

7.3 Identify methods of surface preparation before marking out.

7.4 Identify measuring and marking-out techniques.

7.5 Identify the engineering applications of measuring and marking out.

7.6 Identify the methods of supporting workpieces while measuring or marking out.

7.7 Identify errors that can occur when measuring and marking out.

Chapter summary

There are many stages in the manufacture of an engineering part or component. These include the selection of appropriate materials and processes and the selection of appropriate tools and equipment. You will find plenty of information on how these tasks are performed in the other chapters in the book but now we focus our attention on another important task, namely that of selecting and applying appropriate techniques for marking out the work so that it ensures that the finished component conforms fully with its drawing and specification. Paramount in all of this is the ability to be able to make accurate dimensional measurements and transfer these to a workpiece without error. This chapter shows you how it's done.

Learning outcome 7.1

Identify the sources of information used in engineering to support marking-out activities

Before attempting to mark out your work it will be essential to gather and assemble information from several sources. You will need to start with a specification for the component, which will usually include the material to be used as well as physical dimensions and tolerances. You will also need an accurately dimensioned drawing to refer to. The drawing may also provide additional information such as coatings or finishes to be applied.

Primary information sources

Primary information sources for marking out include specifications, work instructions and detailed drawings of the component or assembly.

Secondary information sources

Secondary sources of information include catalogues from tool suppliers (these can often be a useful source of information), data sheets, tables and charts, and online sources. Such information includes data presented in tabular and/or graphical form such as:

* material properties (such as hardness, brittleness, electrical and/or thermal conductivity, density etc.)

- tables of tensile strength and proof strength
- standard drill sizes
- standard thread sizes and pitches
- tapping and clearance drill sizes
- coordinates for equally spaced holes around a circle
- dimensions of Morse taper sleeves
- conversion factors for length, weight, area, volume in the metric and imperial systems
- trigonometric data such as sine, cosine, tangent and cotangents of angles
- angular measurement conversion from radians to degrees
- temperature measurement conversion from Fahrenheit to Celsius.

The above list is not exhaustive and you may not need to have all of it to hand but, for anything other than the most basic marking-out task, some additional data will almost certainly be required. Before starting work on a particular task, you may need to consider various factors before you can arrive at a comprehensive list of the information that you need. These factors are largely governed by the materials that you will be using and the processes that are going to be applied to them. You should begin by:

- establishing the material(s) to be used
- checking the properties and form of the material that you will be working with
- checking the dimensions, volume and weight (as appropriate) of the finished part
- ensuring that you have the latest version of any drawings and that these are supplied with dimensions, tolerances and clearances (as appropriate)
- examining the drawings and specifications carefully, and checking that there is no ambiguity or discrepancy in them
- selecting the most appropriate equipment and tooling (and checking that it is available)
- selecting and preparing a reference or datum to work from
- calculating relevant dimensions and angles where appropriate
- checking that the profile accurately matches the drawing and specifications.

This list might sound rather daunting but it is logical and, if you can 'check off' each of the items on the list, you should be ready to begin marking out a part or component. Next, we will look at the equipment that you need to actually do the marking out!

Key point

Marking out is the process of laying out and transferring a design or pattern to a workpiece. Following its design, this is the first step in the manufacture of a part or assembly.

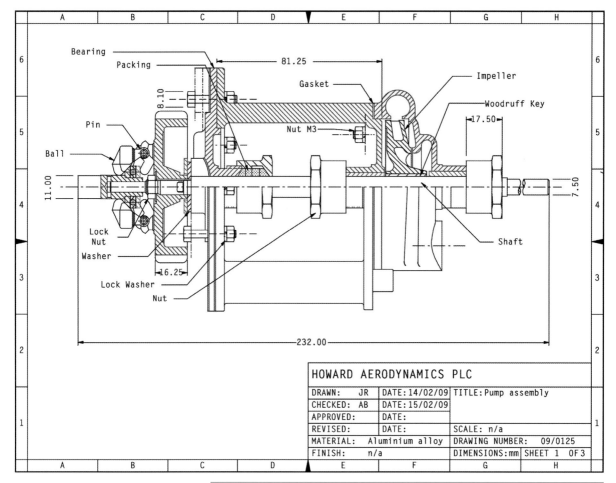

Figure 7.1 A detail drawing will be required prior to marking out. The drawing will usually provide information on dimensions, tolerances, materials and finishes, as appropriate.

Test your knowledge 7.1

Explain why it is essential to have the latest specification and/or drawings of a part or component prior to marking out.

Learning outcome 7.2

Identify the range of measuring and marking-out equipment available

Prior to cutting metal or any other material, several items of equipment will be required for marking out. The items that you need will depend not only on the nature and physical dimensions of the part but also on the form in which the material is supplied. However, in all of this the first skill that you will need to master is that of being able to make accurate measurements. This might sound easier than it is!

Measurement

Before you can mark out a component or check it during manufacture you need to know about engineering measurement. All engineering measurements are *comparative* processes. This basically means that you need to compare the size of the feature with an accurately known standard.

Figure 7.2 shows a steel rule and how to use it. The distance between the lines or the width of the work is being compared with

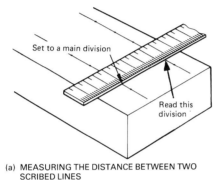

(a) MEASURING THE DISTANCE BETWEEN TWO
     SCRIBED LINES

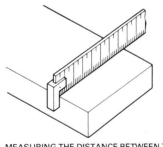

(b) MEASURING THE DISTANCE BETWEEN TWO
     FACES USING A HOOK RULE

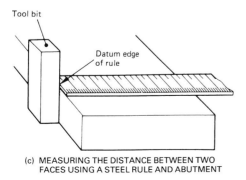

(c) MEASURING THE DISTANCE BETWEEN TWO
     FACES USING A STEEL RULE AND ABUTMENT

Figure 7.2 Using a steel rule

the rule. In this instance, the rule is our standard of length. The rule should be made from spring steel and the markings should be engraved into the surface of the rule. The edges of the rule should be ground so that it can be used as a straight edge, and the datum end of the rule should be protected from damage so that the accuracy of the rule is not lost.

Test your knowledge 7.2

State three precautions that you should take to keep a steel rule in good condition.

Calipers

To increase the usefulness of the steel rule and to improve the accuracy of taking measurements, accessories called calipers are used. These are used to transfer the distances between the faces of the work to the distances between the lines engraved on the rule. Figure 7.3 shows some different types of inside and outside calipers. It also shows how to use them.

A steel rule can only be read to an accuracy of about ±0.5mm. This is rarely accurate enough for precision engineering purposes.

Figure 7.4a shows a vernier caliper and how it can take inside and outside measurements. Typical vernier scales are shown in

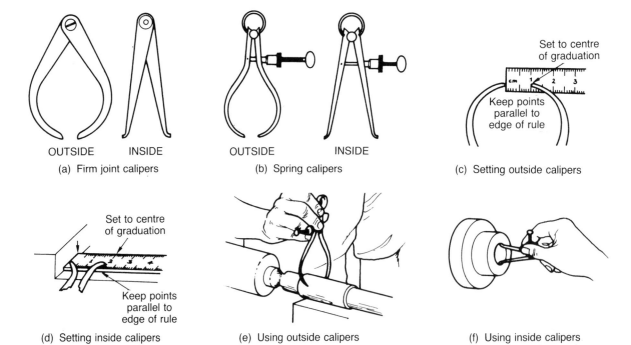

(a) Firm joint calipers

(b) Spring calipers

(c) Setting outside calipers

(d) Setting inside calipers

(e) Using outside calipers

(f) Using inside calipers

Figure 7.3 Inside and outside calipers.

Figure 7.4b. Some verniers have scales that are different to the ones shown in Figure 7.4b. Always check the scales before taking a reading. Like all measuring instruments, a vernier caliper must be treated carefully and it must be cleaned and returned to its case whenever it is not in use.

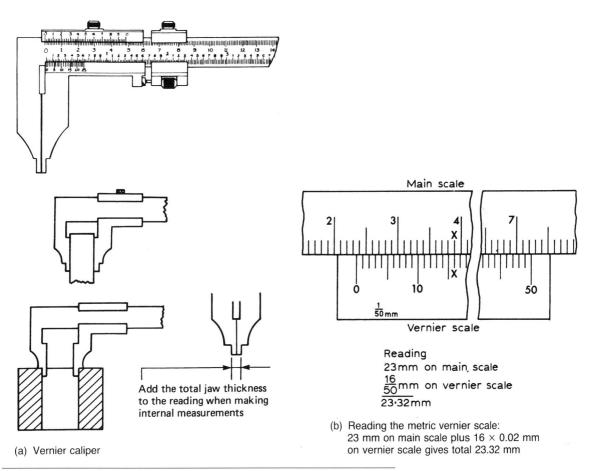

Add the total jaw thickness to the reading when making internal measurements

(a) Vernier caliper

Main scale

Vernier scale

Reading
23 mm on main scale
$\frac{16}{50}$ mm on vernier scale
23·32 mm

(b) Reading the metric vernier scale:
23 mm on main scale plus 16 × 0.02 mm on vernier scale gives total 23.32 mm

Figure 7.4 Vernier calipers.

Vernier calipers are difficult to read accurately even if you have good eyesight. A magnifying glass is helpful. The larger sizes of vernier caliper are quite heavy and it may be difficult to get a correct and consistent 'feel' between the instrument and the work. An alternative instrument is the micrometer caliper. Figure 7.5 shows a micrometer caliper and the method of reading its scales. Nowadays micrometers are available with digital readout. This makes them much easier to use but you should still know how to make appropriate use of a non-digital instrument.

Micrometer calipers are more compact than vernier calipers. They are also easier to use. The ratchet on the end of the thimble ensures that the contact pressure is kept constant and at the correct value.

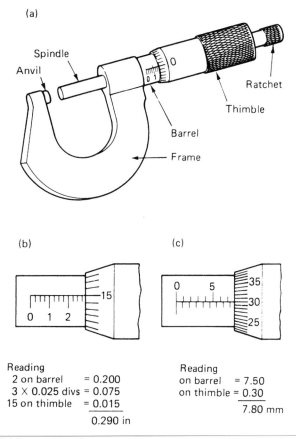

(a)

Spindle

Anvil

Ratchet

Thimble

Barrel

Frame

(b)

(c)

Reading
2 on barrel   = 0.200
3 × 0.025 divs = 0.075
15 on thimble   = 0.015
                  ‾‾‾‾‾‾‾
                  0.290 in

Reading
on barrel   = 7.50
on thimble = 0.30
              ‾‾‾‾‾
              7.80 mm

Figure 7.5 Micrometer caliper.

Unfortunately, micrometer calipers have only a limited measuring range (25mm), so you need a range of micrometers moving up in size in 25mm steps. (0 to 25mm, 25 to 50mm, 50 to 75mm and so on to the largest size). You also need a range of inside micrometers and depth micrometers.

Test your knowledge 7.3

Explain how you would obtain a constant measuring pressure when using a micrometer caliper.

Test your knowledge 7.4

1 Determine the reading of the micrometer caliper scales shown in Figure 7.6a.
2 Determine the reading of the vernier caliper scales shown in Figure 7.6b.

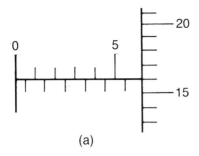

(a)

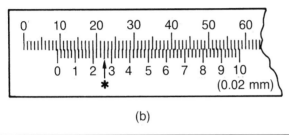

(b)

Figure 7.6 See Test your knowledge 7.4.

Angular measurements

The most frequently measured angle is 90°. This is a *right angle* and surfaces at right angles to each other are said to be perpendicular. Right angles are checked with a *try square*. Figure 7.7 shows a typical engineer's try square and two ways in which it can be used.

Test your knowledge 7.5

Name the instrument that is used to check that two surfaces are at right angles to one another.

For angles other than a right angle, a protractor is used. This may be a simple plain protractor as shown in Figure 7.8a or it may have a vernier scale (vernier protractor) as shown in Figure 7.8b.

Test your knowledge 7.6

State the reading accuracy of the vernier bevel protractor shown in Figure 7.8b.

Test your knowledge 7.7

Explain how the reading stated in Figure 7.8b is obtained.

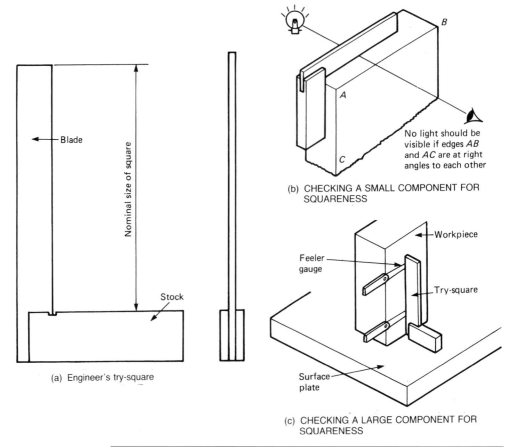

(a) Engineer's try-square

(b) CHECKING A SMALL COMPONENT FOR SQUARENESS

No light should be visible if edges *AB* and *AC* are at right angles to each other

(c) CHECKING A LARGE COMPONENT FOR SQUARENESS

Figure 7.7 An engineer's try square.

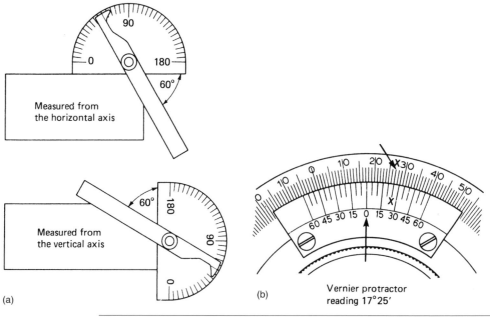

(a)

Measured from the horizontal axis

Measured from the vertical axis

(b)

Vernier protractor reading 17°25′

Figure 7.8 Using a protractor.

Tolerancing and gauging

So far we have only considered measurement of size. This is usual when making a small number of components. However, for quantity production it requires too high a skill level, is too time consuming and, therefore, too expensive. Since no product can be made to an exact size, nor measured exactly, the designer usually gives each dimension an upper and lower size. This is shown in Figure 7.9. If the component lies anywhere between the upper and lower limits of size it will function correctly. The closer the limits, the more accurately the component will work, but the more expensive it will be to make.

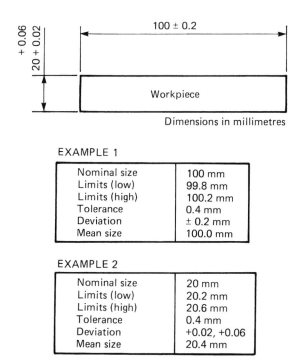

EXAMPLE 1

Nominal size	100 mm
Limits (low)	99.8 mm
Limits (high)	100.2 mm
Tolerance	0.4 mm
Deviation	± 0.2 mm
Mean size	100.0 mm

EXAMPLE 2

Nominal size	20 mm
Limits (low)	20.2 mm
Limits (high)	20.6 mm
Tolerance	0.4 mm
Deviation	+0.02, +0.06
Mean size	20.4 mm

Figure 7.9 Use of tolerances.

A major advantage of using tolerance dimensions is that they can be checked without having to be measured. Gauges can be used instead of measuring instruments. This is easier, quicker and much cheaper. Figure 7.10 shows how a caliper gauge can be used to check the thickness of a component. Plug gauges are used in a similar manner to check hole sizes.

Some more gauges are shown in Figure 7.11. Radius gauges are used to check the corner radii of components. Feeler gauges are used to check the gap between components – for example, the valve tappet clearances in a motor vehicle engine. Thread gauges are used to check the pitch of screw threads.

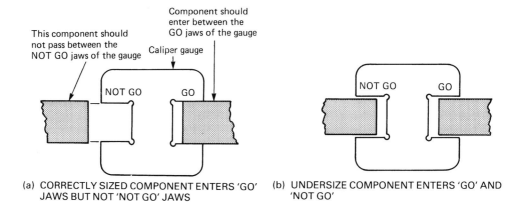

(a) CORRECTLY SIZED COMPONENT ENTERS 'GO' JAWS BUT NOT 'NOT GO' JAWS

(b) UNDERSIZE COMPONENT ENTERS 'GO' AND 'NOT GO'

(c) OVERSIZE COMPONENT DOES NOT ENTER 'GO' OR 'NOT GO'

Figure 7.10 Using a caliper gauge to check the thickness of a component.

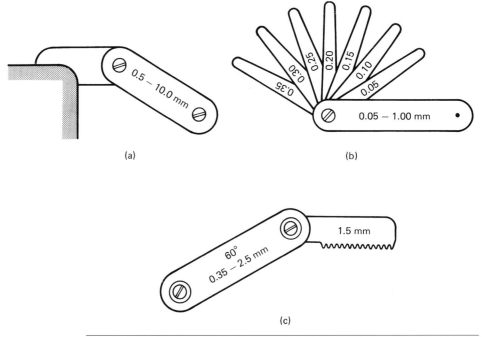

Figure 7.11 Radius and feeler gauges.

Test your knowledge 7.8

With reference to Figure 7.12, state the following:

a)   the nominal size
b)   the upper limit
c)   the lower limit
d)   the tolerance
e)   the deviation
f)   the mean size

Test your knowledge 7.9

With the aid of sketches, explain how you would check the dimensions for the hole shown in Figure 7.12 with a plug gauge.

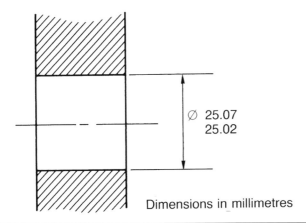

Dimensions in millimetres

Figure 7.12 See Test your knowledge 7.8 and 7.9.

Activity 7.1

Select suitable measuring instruments and explain how you would check the dimensions for the component shown in Figure 7.13. Present your work in the form of a brief written report including sketches where appropriate.

Key point

Tolerancing involves comparing a measurement against upper and lower limits.

Test your knowledge 7.10

Write down the readings for:
a)   the micrometer shown in Figure 7.14a
b)   the vernier shown in Figure 7.14b

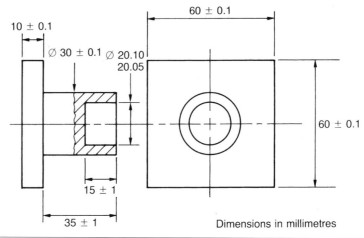

Figure 7.13 See Activity 7.1.

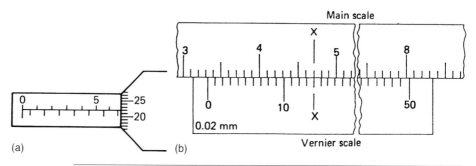

Figure 7.14 See Test your knowledge 7.10.

Learning outcome 7.3

Identify methods of surface preparation before marking out

Before starting to mark out an uncoated metal surface it must be cleaned and all surface deposits such as oil, grease and any loose material that might be present must be carefully removed. Unfortunately, scribed lines and other marks made by punches tend not to contrast with the background. To make them stand out it is often necessary to apply a coloured coating to the surface. This is often done using a blue stain called *marking blue*. This comprises a thin layer of dye that can be easily applied and removed. Scribe and punch markings usually show up very clearly against it. A particular advantage of using marking blue is that, following its application and prior to marking out, any existing marks and scratches will be obliterated so that only the applied scribed lines will later be visible.

An alternative to the use of marking blue is the use of a felt-tip marker. This can often be more convenient than dye coatings and it can be easily applied and removed. Where marking out is to be applied to castings or forgings a layer of 'whitewash' can be applied.

Machined surfaces can be treated in a similar manner but care will be needed to avoid the marking being removed when a part is being handled during the marking-out process. On rough structures, such as castings or forgings, whitewash or a mixture of chalk and water can be used. With welded components tolerances will be less critical and marking out can often be performed with chalk, felt-tip markers and centre punches.

> **Key point**
>
> Prior to marking out it is important to ensure that the surface of the material is cleaned and free from oil, grease and any other surface deposits.

Test your knowledge 7.11

Explain why 'marking blue' is often used to coat a material prior to marking out.

Learning outcome 7.4

Identify measuring and marking-out techniques

Just in case you're wondering, here's why you need to mark out your work:

- To provide you with *guide lines*. These lines mark features on the surface of your work and give you a handy reference to work to, particularly when not using machine controls for setting distances.
- To provide you with a guide when *setting up* your work for more accurate machining where the machine controls will ultimately determine the accuracy of your work.
- To provide you with an indication that will help you confirm that sufficient *machining allowance* has been left when you are working on cast or forged components. This can be important when you need to be sure that features such as webs, flanges and cored holes have been correctly positioned.

Scribed lines and centre marks

Scribed lines are fine lines cut into the surface of the material being marked out by the point of a scribing tool (scriber). An example of a scriber is shown in Figure 7.15a. To ensure that the scribed line shows up clearly, the surface of the material to be marked out is coated with a thin film of a contrasting colour. For example, the

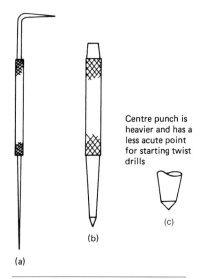

Centre punch is heavier and has a less acute point for starting twist drills

(c)

(b)

(a)

Figure 7.15 Scriber, dot punch and centre punch.

surfaces of the casting to be machined are often whitewashed. Bright metal surfaces can be treated with a marking-out 'ink'. Plain carbon steels can be treated with copper sulphate solution, which copper plates the surface of the metal. This has the advantages of permanence. Care must be taken in its use as it will attack any marking-out and measuring instruments with which the copper sulphate comes into contact.

Centre marks are made with a *dot punch* as shown in Figure 7.15b or with a *centre punch*, as shown in Figure 7.15c. A dot punch has a fine conical point with an included angle of about 60°. A centre punch is heavier and has a less acute point angle of about 90°. It is used for making a centre mark for locating the point of a twist drill and preventing the point from wandering at the start of a cut.

The dot punch is used for two purposes when marking out.

* A scribed line can be protected by a series of centre marks made along the line, as shown in Figure 7.16a. If the line is accidentally removed, it can be replaced by joining up the centre marks. Further, when machining to a line as shown in Figure 7.16b, the half-marks left behind are a witness that the machinist has 'split the line'.

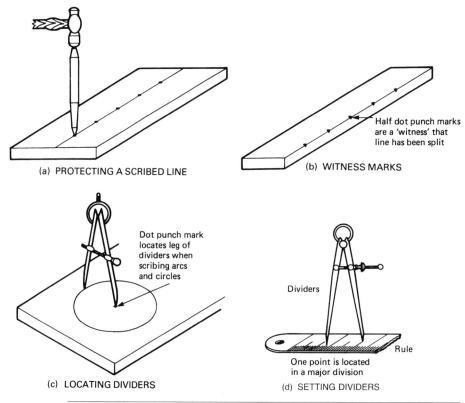

(a) PROTECTING A SCRIBED LINE

Half dot punch marks are a 'witness' that line has been split

(b) WITNESS MARKS

Dot punch mark locates leg of dividers when scribing arcs and circles

(c) LOCATING DIVIDERS

Dividers

Rule

One point is located in a major division

(d) SETTING DIVIDERS

Figure 7.16 Using dot punch and dividers.

- Second, dot punch marks are used to prevent the centre point of dividers from slipping when scribing circles and arcs of circles, as shown in Figure 7.16c. The correct way to set divider points is shown in Figure 7.16d.

It is important to note that, when a centre punch is driven into the work, distortion can occur. This can be a burr raised around the punch mark, swelling of the edge of a component or the buckling of thin material.

Learning outcome 7.5

Identify the engineering applications of measuring and marking out

From what we have already seen, your basic requirements for marking out are:

- a scriber to produce a line
- a rule to measure distances and act as a straight edge to guide the point of the scriber
- dividers to scribe circles and arcs of circles as shown in Figure 7.16c.

In addition, you require hermaphrodite or *odd-leg calipers*, as shown in Figure 7.17a and a try square. Odd-leg calipers are used to scribe lines parallel to a datum edge. A try square and scriber are used to scribe a line at right angles to a datum edge, as shown in Figure 7.17b.

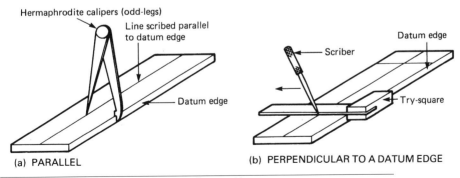

(a) PARALLEL (b) PERPENDICULAR TO A DATUM EDGE

Figure 7.17 Using odd-leg calipers, scriber and try square.

Test your knowledge 7.12

Describe, with the aid of sketches, how you would mark out the link shown in Figure 7.18 on a 6mm-thick low carbon steel.

Test your knowledge 7.13

Refer to Figure 7.18. Explain with the aid of diagrams how you would mark out the centre lines for the holes and radii using a surface table and scribing block, together with whatever additional equipment is required.

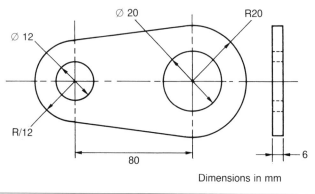

Dimensions in mm

Figure 7.18 See Test your knowledge 7.12 and 7.13.

Alternatively, lines can be scribed parallel to a *datum edge* using a *surface table* and *scribing block*, as shown in Figure 7.19a. In this example the scribing point is set to a steel rule, so the

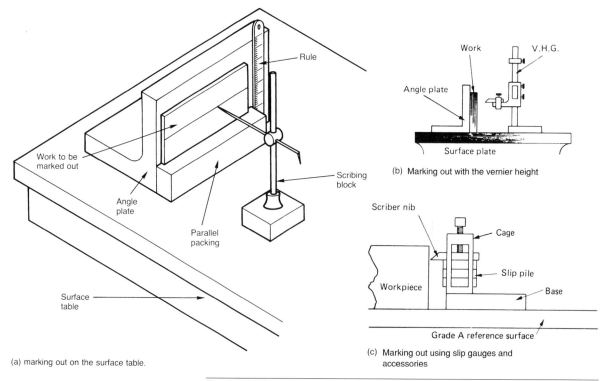

(a) marking out on the surface table.

(b) Marking out with the vernier height

(c) Marking out using slip gauges and accessories

Figure 7.19 Marking out using a surface table.

accuracy is limited. Alternatively, the line can be scribed using a *vernier height gauge*, as shown in Figure 7.19b. This is very much more accurate. Where extreme accuracy is required, *slip gauges* and slip gauge accessories can be used, as shown in Figure 7.19c.

Cylindrical components are difficult to mark out since they tend to roll about. To prevent this they can be supported on *vee-blocks*, as shown in Figure 7.20a. Vee-blocks are always made and sold in boxed sets of two. In order that the axis of the work is parallel to the surface plate or table, you must always use such a matched pair of vee-blocks, and make sure that, after use, they are put away as a pair.

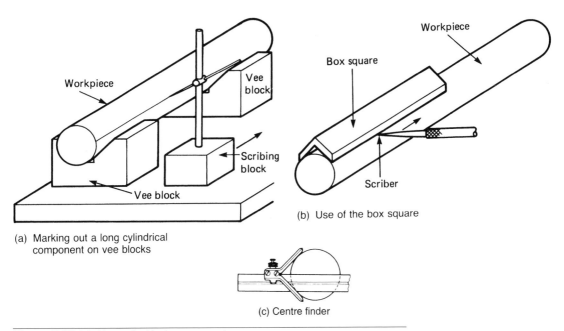

(a) Marking out a long cylindrical component on vee blocks

(b) Use of the box square

(c) Centre finder

Figure 7.20 Marking cylindrical components.

Another useful device for use on cylindrical work is the *box square* shown in Figure 7.20b. This is used for scribing lines along cylindrical work parallel to the axis. A *centre finder* is shown in Figure 7.20c. This is used to scribe lines that pass through the centre of circular blanks or the ends of cylindrical components.

Key point

Tools used for marking out include scribers, rules, dividers and calipers.

Test your knowledge 7.14

With the aid of sketches, explain how you would mark out a keyway 8mm wide by 4mm deep and 50mm long on the end of a 38mm diameter shaft.

Datum points, lines and surfaces

First, we had better revise what we know about rectangular and polar coordinates. Examples of these are shown in Figure 7.21.

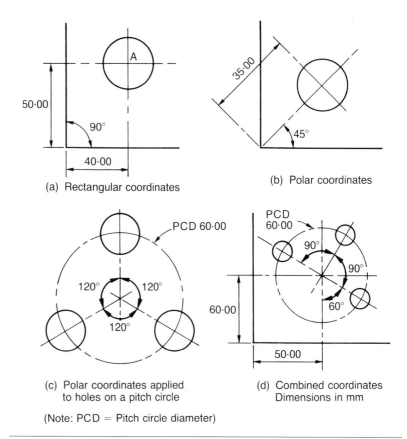

(a) Rectangular coordinates

(b) Polar coordinates

(c) Polar coordinates applied to holes on a pitch circle

(d) Combined coordinates Dimensions in mm

(Note: PCD = Pitch circle diameter)

Figure 7.21 Datum points and coordinates.

Rectangular coordinates

The point A in Figure 7.21a is positioned by a pair of ordinates (coordinates) lying at right angles to each other. They also lie at right angles to the datum edges from which they are measured. This system of measurement requires the production of two datum surfaces or edges at right angles to each other. That is, two datum edges that are mutually perpendicular.

Polar coordinates

Polar coordinates consist of one linear distance and an angle, as shown in Figure 7.21b. Dimensioning in this way is useful when the work is to be machined on a rotary table. It is widely used when setting out hole centres round a pitch circle as shown in Figure 7.12c. Quite frequently both systems are used at the same time, as shown in Figure 7.21d.

Test your knowledge 7.15

Draw and dimension eight holes each of 10mm diameter on a 75mm diameter pitch circle. The pitch circle is to be central on a plate measuring 100mm by 125mm.

Datum points

During our discussion on marking out, we have kept referring to datum points, datum lines and datum edges. A datum is any point, line or surface that can be used as a basis for measurement. When you go for a medical check-up, your height is measured from the floor on which you are standing. In this example the floor is the basis of measurement, it is the datum surface from which your height is measured.

Point datum

This is a single point used as a reference from which a number of features are marked out. For example, Figure 7.22a shows two concentric circles representing the inside and outside diameter of a pipe-flange ring. It also shows the pitch circle around which the bolt hole centres are marked off. All these are marked out using dividers or trammels (beam compasses) from a single-point datum.

Line datum

Any line from which, or along which, a number of features are marked out. An example of the use of line datums is shown in Figure 7.22b.

Surface datum

This is also known as an *edge datum* or *service edge*. It is the most widely used datum for marking out solid objects. Two edges are accurately machined at right angles to each other and all the dimensions are taken from these edges. Figure 7.22c shows how dimensions are taken from surface datums. It shows both rectangular and polar coordinates. Alternatively, the work can be clamped to an angle-plate. The mutually perpendicular edges of the angle-plate provide the surface datums. In this instance there is no need to machine the edges of the workpiece at right angles.

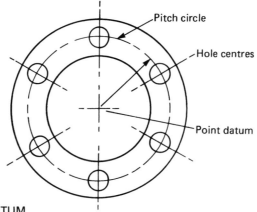

(a) SINGLE-POINT DATUM

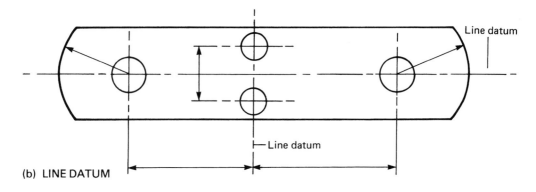

(b) LINE DATUM

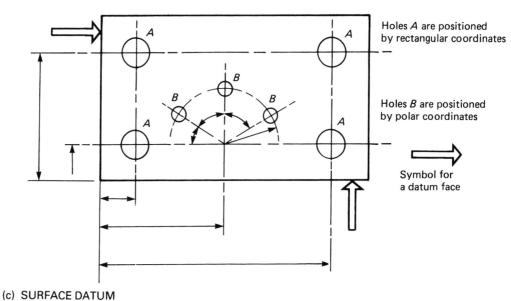

(c) SURFACE DATUM

Figure 7.22 Point, line and surface datums.

Learning outcome 7.6

Identify the methods of supporting workpieces while measuring or marking out

You have already seen how various different methods are used to support a material or component when it is being marked out. These include:

- *Surface plates*. These provide a true flat surface on which a part or component can be placed, which can be used with try squares, scribing blocks and height-measuring instruments (see Figure 7.19).
- *Vee blocks*. These provide a means of supporting cylindrical components. They are usually placed on a surface table and used in conjunction with scribing blocks and height-measuring instruments (see Figure 7.20).
- *Box squares*. These allow lines to be accurately made parallel with the central axis of cylindrical components (see Figure 7.20).
- *Centre finders*. These provide a neat and effective way of ensuring that a cylindrical component is aligned when marking its centre (see Figure 7.20).

Surface plates

Surface plates are used to provide the baseline reference for measurements made on a workpiece. Because of this, the surface of the plate must be extremely flat, and must not be liable to movement or vibration when measurements are being carried out. Surface plates must also be large enough to support a workpiece together with appropriate measuring and marking-out instruments such as scribers, squares and height-measuring instruments. Surface plates vary in size from 160mm × 100mm to 2.5m × 1.6m. The larger surface plates can be extremely heavy and should be treated with great care when being moved.

Most surface plates are made from cast iron with an accurately machined table surface. Cast iron is strong but susceptible to corrosion and rust. An alternative to the use of cast iron is granite. Granite surface plates must be treated with care since they are susceptible to being chipped. This does not usually affect the accuracy of the overall plane since, even when chipped, there can be sufficient flat surface remaining in order to ensure that the workpiece sits on a true surface. With cast iron plates, surface contamination is likely to raise the surface of the plate at the point at which it occurs. In such cases the workpiece may no longer be

Key point

Surface plates are used to provide the baseline reference for measurements made on a workpiece.

resting in a truly flat position. Granite also offers the advantage that, unlike cast iron, it is non-magnetic and also will not rust.

Test your knowledge 7.16

Give two advantages of using granite when compared with cast iron for the manufacture of a surface plate.

Test your knowledge 7.17

Describe three different methods of holding a workpiece during marking out. Give an example of the type of component part that would be appropriate for each of the methods that you describe.

Learning outcome 7.7

Identify errors that can occur when measuring and marking out

Earlier we mentioned the need to ensure that the marking-out process is sufficiently rigorous to ensure that the dimensions of the finished part or component part will be within the stated tolerance limits. To ensure that this is the case, accuracy and consistency are essential. *Accuracy* relates to the quality of the measurements and how close the marked-out dimensions are to those given in specifications and drawings. *Consistency* relates to the ability to maintain the same amount of accuracy across all of the dimensions. Where dimensional values lie outside the stated tolerance limits, errors will need to be accounted for.

> **Key point**
>
> A tolerance error is the difference between the dimension on the engineering drawing and the measurement of the manufactured part or component.

Tolerance errors

The difference between the dimension given in a specification or shown on a drawing and the measurement of the manufactured part or component is referred to as a tolerance error. Such errors are often 'one-off', for example, occurring as a result of misreading an instrument scale or marking on the wrong side of a rule. Fortunately, 'one-off' errors are usually easy both to spot and avoid using simple checks and measurements against the original drawings and specifications.

> **Key point**
>
> A cumulative error occurs when the tolerance errors of different points and features are added together.

Cumulative errors

Cumulative errors may sometimes occur – for example, if a series of holes are marked out using the distance measured from the centre of one to the centre of the next rather than marking the centres using the distance from the same datum point. An error measuring the distance will not only mean that all of the other centres are incorrectly marked but also that the error will increase with the number of centres marked.

Key point

A datum is a point, line or surface from which dimensions are taken. A datum point can help eliminate cumulative errors and so its use can help to improve accuracy.

Test your knowledge 7.18

Explain how a cumulative error differs from a tolerance error.

Review questions

1. Identify **four** sources of information that are commonly needed in conjunction with marking-out activities. In what typical circumstances would each of the sources that you've listed be used?

2. Sketch an engineer's try square. State what it is used for and explain how it is used.

3. With the aid of a sketch, distinguish between inside and outside calipers and explain what they are used for.

4. Identify **two** methods of surface preparation before marking out.

5. Explain why it is necessary to ensure that a workpiece is clean and free from oil and grease before marking out.

6. Describe **two** different methods of holding a workpiece while marking out. Give an example of where each method is used.

7. State **two** disadvantages of using a cast iron surface plate when compared with a granite surface plate.

8. With the aid of sketches, show how the location of the centre of a circular hole in a flat plate can be specified from a fixed datum point using:
   a) rectangular coordinates
   b) polar coordinates

9. A workpiece has a nominal size of 120mm. If a deviation of 0.3mm is allowed above the nominal size and 0.5mm below the nominal size, what is the tolerance of the workpiece dimension and what will its upper and lower tolerance limits be?

10. When marking out, explain why the use of a datum point can help avoid cumulative errors.

Chapter checklist

Learning outcome	Page number
7.1 Identify the sources of information used in engineering to support marking-out activities.	184
7.2 Identify the range of measuring and marking-out equipment available.	186
7.3 Identify methods of surface preparation before marking out.	196
7.4 Identify measuring and marking-out techniques.	197
7.5 Identify the engineering applications of measuring and marking out.	199
7.6 Identify the methods of supporting workpieces while measuring or marking out.	205
7.7 Identify errors that can occur when measuring and marking out.	206

Engineering mathematics and science principles

CHAPTER **8**

Engineering materials

Learning outcomes

When you have completed this chapter you should understand how to select engineering materials, including being able to:

8.1 Select a basic engineering material suitable for a given application.

8.2 Recognize the classification, range and application of materials used in engineering.

8.3 State the forms of supply of engineering materials.

8.4 Identify the factors that make materials suitable for engineering applications.

Chapter summary

Welcome to the fascinating world of engineering materials! All branches of engineering are concerned with the use and processing of materials and so this unit makes an excellent companion to the others that you study as part of your EAL Level 2 award. In fact, the correct choice and processing of materials can be crucial. For example, civil engineers need to ensure that the materials they use to build roads, bridges and dams are able to withstand the forces and stresses that will be placed on them. They also need to be confident that the building materials will not degrade with time.

Think about what might happen if the materials used were not fit for purpose. For example, if a suspension bridge was built from rope rather than steel cable it might not be able to support a heavy load. It would also degrade with time. Even on a much smaller scale, knowledge of materials is vitally important. An electronic engineer designing a complex flight control system must ensure that the materials chosen do not deteriorate in the harsh environment in which the system will operate. Once again, failure could be catastrophic, particularly if the failure just happened to occur at a critical point in flight, such as take-off or landing. A study of materials is, therefore, an essential part of every engineer's portfolio and for this reason alone, is worthy of your attention.

Learning outcome 8.1

Select a basic engineering material suitable for a given application

This unit is about the materials used in engineering so let's start by looking at the materials used in a typical engineered product. Figure 8.1 shows the interior of a low-voltage d.c. power supply of the type used to supply power to small items of consumer electronic equipment. The power supply is rated at 12V, 1A maximum. For safety reasons, the power supply is totally enclosed and connection to the a.c. mains supply is made by means of cable fitted with a standard fused 13A mains plug. Connection to the equipment being powered is made by a second power lead fitted with a d.c. power connector.

The main components used in the low-voltage d.c. power supply have been labelled A to H in Figure 8.1. They are as follows:

Enclosure: A

The enclosure is made from a plastic material known as acrylonitrile butadiene styrene (ABS). ABS is tough and resists impact. The enclosure not only provides protection for the components inside the power supply but it also reduces the risk of damage and electric shock that might result from contact with 'live' circuitry. The plastic material is chosen because it provides insulation, is inexpensive, and can be moulded into a complex shape (note how the supporting pillars and ventilation slots are integral parts of the case).

Transformer core: B

The transformer core is made from laminated strips of steel that are stamped from thin sheets of metal. Steel is chosen as the material for the transformer core because of its excellent ferromagnetic properties (see Chapter 11).

Transformer windings: C

The coils that make up the transformer windings (not easily seen in Figure 8.1) are made from enamelled copper wire. Copper is used because it has a low resistance and is non-magnetic. The wire is drawn (or pulled out) in a continuous length of the required wire diameter and then given a thin enamel surface coating in order to provide insulation. The insulation is needed in order to avoid the risk of shorts between adjacent turns of wire.

Power lead: D

The power lead uses copper wire which has multiple strands (to make it flexible). Once again, copper is chosen because it is an excellent conductor. The stranded copper wire is given a plastic insulating coating.

Printed circuit board: E

The printed circuit board is made from a composite material (glass-fibre reinforced plastic). This material is strong and is a good electrical insulator. The material is given a copper laminating coating which is etched in order to provide the required conducting track pattern which links the component leads and pins together.

Heat dissipator: F

The heat dissipator (or 'heat sink') is designed to conduct heat away from an integrated circuit voltage regulator (not visible in Figure 8.1) and convect it into the surrounding space. The heat dissipator is made from aluminium alloy (which is a good conductor of heat and is also relatively light). The complex finned shape is made from an extrusion in which hot alloy is forced under pressure into a shape which has a continuous cross-section.

Semiconductors: G

Like most electronic gadgets, the power supply uses several different types of semiconductor device. These include four rectifier diodes, a light-emitting diode and an integrated circuit voltage regulator. These devices are based on semiconductor materials, a class of material that is neither a conductor nor an insulator. The ability of a semiconductor device to carry an electric current depends on the amount and polarity of the supplied voltage. The materials used in the manufacture of the semiconductors in the power supply include silicon and gallium arsenide.

Retaining screws: H

The five pan-head retaining screws are made from steel with a black oxide finish. The screws are designed to form their own threads when inserted into the pillars in the (much softer) plastic case.

Finally, it's important to note that the materials used for these components have different properties and they require quite different treatments and manufacturing processes. Being able to select appropriate materials and specify the correct treatments and manufacturing processes to be used with them is an important skill that you must develop along the road to becoming an engineer.

Key point

ABS is a 'plastic' material that is tough and impact resistant. It can easily be moulded into complicated shapes and acts as an electrical insulator.

Key point

When a material is moulded it is first heated and then fed into a mould. After cooling, the material takes on the shape of the mould.

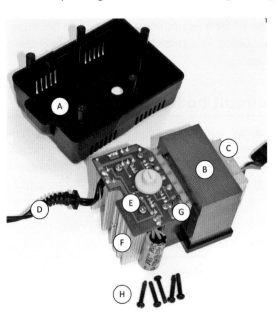

Figure 8.1 A low-voltage d.c. power supply showing some of the main components.

Test your knowledge 8.1

In Figure 8.1, which components are:

a) moulded
b) stamped
c) extruded?

Test your knowledge 8.2

In Figure 8.1, explain why copper is used in the manufacture of the parts labelled C, D and E.

Test your knowledge 8.3

Give **three** reasons for using ABS to manufacture the power supply enclosure shown in Figure 8.1.

Learning outcome 8.2

Recognize the classification, range and application of materials used in engineering

By now, you should be beginning to appreciate that a very wide range of materials, parts and components are used in engineering! You should also be aware that a number of different processes can be applied in order to turn a raw material into a finished product. These processes include cutting, stamping, machining, extruding and moulding. Figure 8.2 shows the main groups of engineering materials.

When selecting a material, you need to make certain that it has properties that are appropriate for the job it has to do. For example, you should ask yourself the following questions:

- Will the material degrade or corrode in its working environment?
- Will the material weaken when it gets hot?
- Will the material break under normal working conditions?
- Can the material be easily cast, formed or cut to shape?

You can assess the suitability of different materials for a particular engineering application by comparing their properties, the forms in which they are supplied and the processes that can be applied to them but, before going into more detail, it is worth attempting to classifying materials into groups of related materials. Members

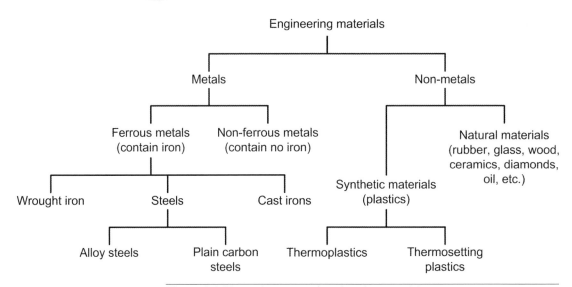

Figure 8.2 The main groups of engineering materials.

of a particular class of material are likely to share a common set of properties (note that this is not always the case!) and that makes them suitable for use in a similar range of applications.

Classes of material

To aid your understanding of the different types of materials and their properties I have divided them into four classes: metals, polymers, ceramics and composites. Strictly speaking, composites are not really a separate group since they are made up from the other categories of material. However, because they display unique properties and are a very important engineering group, they are treated separately here.

Metals and metal alloys are widely used in engineering. The Latin name for iron is *ferrum,* so it is not surprising that ferrous metals and alloys are all based on the metal iron. Alloys consist of two or more metals (or metals and non-metals) that have been brought together as compounds or solid solutions to produce a metallic material with special properties. For example, an alloy of iron, carbon, nickel and chromium is *stainless steel.* This is a corrosion-resistant ferrous alloy. Non-ferrous metals and alloys are the rest of the metallic materials available. Non-metals can be natural, such as rubber, or they can be synthetic such as the plastic compound PVC.

Metals

You will be familiar with metals such as aluminium, iron and copper, in a wide variety of everyday applications, i.e. aluminium saucepans, copper water pipes and iron stoves. Metals can be mixed with other elements (often other metals) to form an alloy. Metal alloys are used to provide improved properties because they are often stronger or tougher than the parent pure metal.

Other improvements can be made to metal alloys by heat treating them as part of the manufacturing process. For example, a very hard steel can be produced by first heating a carbon steel alloy and then rapidly cooling it by immersing it in oil or brine (this process is referred to as 'quenching'). We will be looking at this again later in Chapter 9.

Polymers

Polymers are characterized by their ability to be (initially at least) moulded into shape. They are chemical materials and often have long and unattractive chemical names. There is considerable incentive to seek more convenient names and abbreviations for everyday use. Thus you will be familiar with PVC (polyvinyl chloride) and PTFE (polytetrafluoroethylene). Polymers are made from molecules which join together to form long chains in a process known as polymerization. Teflon is a form of PTFE.

There are several main types of polymer. *Thermoplastics* have the ability to be remoulded and reheated after manufacture. *Thermosetting plastics*, once manufactured, remain in their original moulded form and cannot be reworked. *Elastomers* or rubbers often have very large elastic strains – elastic bands and car tyres are two familiar forms of rubber.

Ceramics

This class of material is again a chemical compound, formed from oxides such as silica (sand), sodium and calcium, as well as silicates such as clay. Glass is an example of a ceramic material, with its main constituent being silica. The oxides and silicates mentioned above have very high melting temperatures and on their own are very difficult to form. They are usually made more manageable by mixing them in powder form, with water, and then hardening them by heating. Ceramics include, brick, earthenware, pots, clay, glasses and refractory (furnace) materials. Ceramics are usually hard and brittle, good electrical and thermal insulators and have good resistance to chemical attack.

Composites

A composite is a material with two or more distinct constituents. These separate constituents act together to give the necessary strength and stiffness to the composite material. The most common example of a composite material today is that of fibre reinforcement of a resin matrix but the term can also be applied to other materials such as metal-skinned honeycomb panels. The property that these materials have in common is that they are light, stiff and strong and, what's more, they can be extremely tough.

Reinforced concrete is another example of a composite material that is invaluable in engineering. The steel and concrete retain their individual identities in the finished structure. However, because they work together, the steel carries the tension loads and the concrete carries the compression loads. Furthermore, although not considered as a separate class of material, some natural materials exist in the form of a composite. The best-known examples are wood, shells and bone. Wood is an interesting example of a natural fibre composite; the longitudinal hollow cells of wood are made up of layers of spirally wound cellulose fibres with varying spiral angle, bonded together with lignin during the growing of the tree.

> **Key point**
>
> Alloys are a mixture of two or more elements, at least one of which is a metal. Alloys often have properties that are different and often superior to the metals they are made from. For example, stainless steel is an alloy of iron, carbon, nickel and chromium.

Test your knowledge 8.4

What are composite materials and how do they differ from metals and ceramics?

Test your knowledge 8.5

What is the important difference between thermoplastics and thermosetting plastics and how does this impact on what they can be used for?

Test your knowledge 8.6

You need a material to line the inside of a furnace. What class of material would be suitable and why?

Activity 8.1

Figure 8.3 shows an off-road vehicle. Use the Internet to search for information on vehicles of this type and use this to complete Table 8.1.

Figure 8.3 Off-road vehicle – see Activity 8.1.

Table 8.1 See Activity 8.1.

Part reference	Part name or function	Material(s) used in manufacture	Reason(s) for using this material
A	Frame		
B	Front fairing		
C	Driver's helmet		
D	Tyres		
E	Shock absorber		
F	Engine block		
G	Wheel rims		

Learning outcome 8.3

State the forms of supply of engineering materials

Engineering materials are supplied in a wide range of different forms including:

- sheet
- plate
- bar
- wire

Key point

A number of different processes can be applied in order to turn a raw material into a finished product. These processes include cutting, stamping, machining, extruding and moulding. It is important to ensure that the form in which the material is supplied is suitable for the processes that will be used during manufacture.

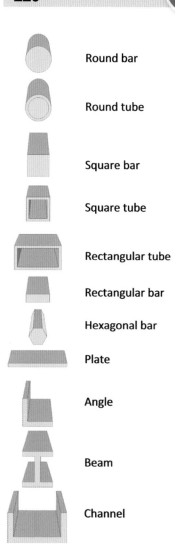

Round bar

Round tube

Square bar

Square tube

Rectangular tube

Rectangular bar

Hexagonal bar

Plate

Angle

Beam

Channel

Figure 8.4 Various forms in which metals and alloys are supplied.

- section
- extrusions
- castings
- wrought
- forgings
- pipe and tube
- hot and cold rolled.

The material used to manufacture a part or component needs to be supplied in a form that makes it possible to use. For example, a rectangular metal component can be relatively easily manufactured from plate or bar but it might prove to be very difficult to manufacture from cylindrical bar. It is thus very important to think about the suitability of a particular form of supply in order to ensure that it will be suitable for the processes that will be needed to turn it into a finished part or component.

As far as possible the chosen form of supply should match the shape and form of the finished work. Because of this it can often be more convenient to use extruded sections and extrusions rather than to machine material from stock bar. For a similar reason, tube stock may often be preferred to cylindrical bar where axial holes need to be bored throughout the length of a cylindrical part or component. Figure 8.4 shows various forms in which metals are supplied (they may not all be available from a particular supplier).

Test your knowledge 8.7

Describe, with the aid of sketches, four different forms in which engineering materials can be supplied.

Activity 8.2

Visit at least three websites of companies that supply steel for engineering manufacture. List at least three different forms and grades of material that each company supplies.

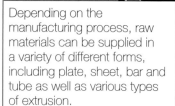

Key point

Depending on the manufacturing process, raw materials can be supplied in a variety of different forms, including plate, sheet, bar and tube as well as various types of extrusion.

Learning outcome 8.4

Identify the factors that make materials suitable for engineering applications

It should be clear that different materials have different characteristics and properties that may make a material suitable for some applications but not others. When selecting a material for a particular application you need to consider a number of factors including the required properties, the cost and (as you've just seen) the form in which the material is supplied. Table 8.2 provides you with some typical examples of selecting materials for some particular engineering applications. When you've looked at Table 8.2 move on to Activity 8.3.

Table 8.2 Typical materials and reasons for their choice for some typical engineering applications.

Application	Chosen material	Reasons for choice
Lathe bed	Cast iron	1. Easy to cast into a complex shape 2. Strong 3. Heavy
Large aircraft body	Aluminium	1. Available in sheet form 2. Easy to cut and shape 3. Light
Electrical cable	Copper	1. Easily drawn into wire/cable 2. Good electrical conductor 3. Corrosion resistant
Speed boat hull	Glass fibre composite	1. Can be moulded into streamline shape 2. Lightweight 3. Corrosion resistant
Machine guard cover	Perspex	1. Transparent 2. Light 3. Can be bent or shaped

Activity 8.3

Brief descriptions of five materials are as follows:

Material A: Supplied in sheets and in various colours; a good electrical insulator with excellent thermal forming properties and good impact strength; ideal for use in the manufacture of cases and enclosures.

Material B: A rather expensive material; has very low friction and adhesion properties; suitable for use over a very wide temperature range; can be extruded or supplied as a coating; has excellent electrical properties.

Material C: A thermoplastic material with good sliding properties; easily bonded, welded and machined; resistant to oils and grease; suitable for use in the manufacture of gears, pulleys and machine parts.

Material D: A rather weak material supplied in opaque clear sheets or as a film; has very high electrical resistance and is ideal for use as an insulator.

Material E: Supplied in clear sheets; a very rigid material; ideal for use as a cover for light fittings; windows and door panels.

1 Identify materials A to E by matching them to the following list: nylon, ABS, polycarbonate, PTFE, polythene.

2 Which materials would be suitable for the manufacture of:

a) a roller bearing

b) a damp-proof membrane

c) a 'non-stick' coating for the plates of a sandwich toaster?

Review questions

1. Explain the difference between ferrous and non-ferrous metals. Give an example of **each** type of metal.

2. Distinguish between a) extrusion and b) moulding of material. Describe **one** example of the use of each process.

3. List **four** classes of material and give an example of a material from each of the classes.

4. Explain what is meant by a polymer material. What makes this class of material different from other classes of material?

5. Give **two** advantages of using glass reinforced plastics (GRP) for the manufacture of vehicle body parts when compared with metals.

6. List **four** different forms in which metals are supplied. Give an example of where each of these forms of supply might be used.

7. List materials that would be appropriate for use in the manufacture of a) a bicycle frame, b) the body of a spark plug, c) a motorcycle fairing, d) the outer casing of a hand-held electric drill. Give reasons for your choice.

Chapter checklist

Learning outcome	Page number
8.1 Select a basic engineering material suitable for a given application.	212
8.2 Recognize the classification, range and application of materials used in engineering.	215
8.3 State the forms of supply of engineering materials.	219
8.4 Identify the factors that make materials suitable for engineering applications.	220

CHAPTER 9

Material properties and heat treatment processes

Chapter summary

We've already mentioned that the materials used in engineering must be fit for purpose and suitable for the application in which they are used. Crucial in all of this are the physical properties of the materials in question. In this chapter we will continue our study of materials by looking in more detail at their properties and the processes that can be applied to them in order to change them. We start by introducing the chemical, electrical and magnetic properties of materials and then move on to the essential physical properties of materials. This chapter also introduces you to a valuable source of reference information in the form of a database of materials and their properties. As you work through the chapter you will be able to add to the database and modify it for your own use.

Learning outcome 9.1

State the properties associated with basic engineering materials

For most engineering applications the most important criterion for the selection of materials is that they do the job properly (i.e. that they perform according to the specification that has been agreed) and that they do it as cheaply as possible. Whether a material does its job depends on its *properties,* which are a measure of how the material reacts to the various influences to which it is exposed. The properties that we might be concerned about as engineers include strength, toughness, hardness, elasticity, rigidity, thermal conductivity, electrical conductivity, and corrosion resistance. Other factors might also be important including density, weight, flexibility, thermal expansion and frictional coefficient.

Activity 9.1

As you progress with your studies you will find it extremely useful to have data on a variety of materials that are commonly used in engineering. Sources of data include reference books and textbooks as well as several online databases accessible via the Internet. You are advised to become familiar with several of these sources. However, in order to provide you with a simplified set of data, I have created the matSdata database specifically for engineering students. The database will provide you with data and data sheets on a wide range of commonly used engineering materials. Furthermore, unlike other databases, matSdata allows you to maintain your own personal copy of the database. This means that you can modify existing records and add your own data as you progress through the course.

You can download a copy of the matSdata database from www.key2engtech.com/matsdata.htm. To get started you should first download the database and then extract the files in order to make your own personal copy in a folder on your hard drive, USB memory stick or SD card. To get you started take a look at the entry for on stainless steel.

Use this information to answer the following questions:

1 Describe the appearance of the material.
2 What are the constituents of the material?
3 State **two** notable properties of the material.
4 What is the density of the material?
5 What is the melting point of the material?
6 What is the appearance of the material?
7 How does the tensile strength of stainless steel compare with other forms of steel?
8 State **two** applications for stainless steel.

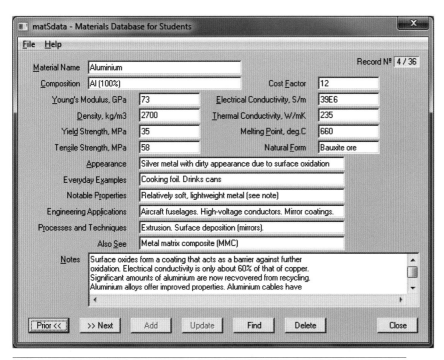

Figure 9.1 The matSdata record for aluminium. You can download your own copy of matSdata from www.key2engtech.com/matsdata.htm

Activity 9.2

Use the matSdata database to locate data sheet records on aluminium and its alloys (i.e. pure aluminium, aluminium 7005 alloy and aluminium-scandium alloy). What is the composition of each of these materials and how do they compare in terms of a) density, b) yield strength and c) tensile strength? Which of the three materials would be most suitable for use in the manufacture of a retaining strut used in the undercarriage of a light aircraft?

What you may have discovered from Activity 9.2 is that materials, even within the same general class, can have very different properties. We shall continue this theme by looking at the different properties that govern the choice of material for a particular engineering application. In the next section we will concentrate on physical properties but for now we will look at chemical, electrical and magnetic properties.

Chemical properties

When engineers look at the chemical properties of a material they are usually concerned with two things: *corrosion* and *degradation*.

Corrosion

This is caused by the metals and metal alloys being attacked and eaten away by chemical substances. For example, the rusting of ferrous metals and alloys is caused by the action of atmospheric oxygen in the presence of water. Another example is the attack on aluminium and some of its alloys by strong alkali solutions.

Degradation

Non-metallic materials do not corrode but they can be attacked by chemical substances. Since this weakens or even destroys the material it is referred to as *degradation*. Unless specially compounded, rubber is attacked by prolonged exposure to oil. Synthetic (plastic) materials can be softened by chemical solvents. Exposure to the ultraviolet rays of sunlight can weaken (perish) rubbers and plastics unless they contain compounds that filter out such rays.

Electrical properties

The electrical properties of a material are important when selecting it for an engineering application that requires the material to either allow or prevent the flow of an electric current to some degree. We specify this property by referring to its electrical resistance.

Electrical resistance

Materials with a very low resistance to the flow of an electric current are good electrical conductors. Conversely, materials with a very high resistance to the flow of electric current are good insulators. Generally, metals are good conductors and non-metals are good insulators (poor conductors). A notable exception is carbon, which conducts electricity despite being a non-metal. The electrical resistance of a metal conductor depends upon:

- length (the longer it is the greater its resistance)
- cross-sectional area (the thicker it is the lower its resistance)
- temperature (the higher its temperature the greater is its resistance)
- resistivity (this is the resistance measured between the opposite faces of a metre cube of the material).

Note that a small number of non-metallic materials, such as silicon, have atomic structures that fall between those of electrical conductors and insulators. These materials are called semiconductors and are used for making solid state devices such as diodes and transistors.

Test your knowledge 9.1

Explain the difference between corrosion and chemical degradation of materials.

Test your knowledge 9.2

List the four factors that determine the electrical resistance of a conductor.

Magnetic properties

All materials respond to strong magnetic fields to some extent, but only the ferromagnetic materials respond sufficiently to be of interest. The more important ferromagnetic materials are the metals iron, nickel and cobalt. Magnetic materials are often divided into two types: soft and hard (this has nothing to do with their physical hardness, which we will meet in the next section).

Soft magnetic materials

Soft magnetic materials, such as soft iron, can be magnetized by placing them in a magnetic field. They cease to be magnetized as soon as the field is removed. Soft magnetic materials can be made more efficient by adding silicon or nickel to the pure iron. Silicon-iron alloys are used for the rotor and stator cores of electric motors and generators. Silicon-iron alloys are also used for the cores of power transformers.

Hard magnetic materials

Hard magnetic materials, such as high carbon steel that has been hardened by cooling it rapidly (quenching) from red heat, also become magnetized when placed in a magnetic field. Hard magnetic materials retain their magnetism when the field is removed. They then become *permanent magnets*. Permanent magnets can be made more powerful for a given size by adding cobalt to the steel to make an alloy.

Test your knowledge 9.3

Explain the difference between soft and hard magnetic materials.

Learning outcome 9.2

Recognize the physical properties of engineering materials

We shall now move on to look at the physical properties of engineering materials. Here we are interested in factors such as strength, toughness and hardness.

Physical properties

Strength

This is the ability of a material to resist an applied force (load) without fracturing (breaking). It is also the ability of a material not to yield. Yielding is when the material elongates suddenly under load and changes shape permanently but does not actually break. This is what happens when metal is bent or folded to shape. The load or force can be applied in various ways, as shown in Figure 9.2.

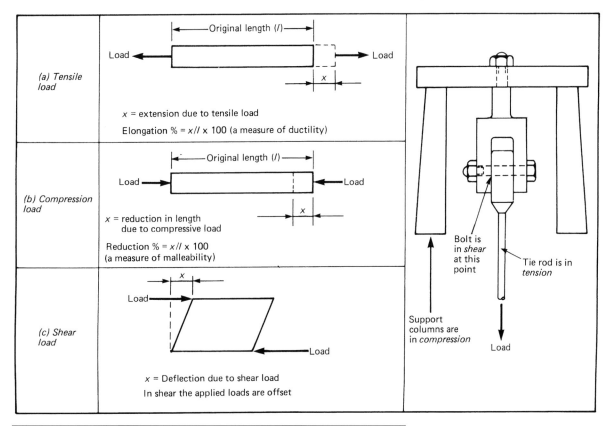

Figure 9.2 Different ways in which a load can be applied.

Toughness

This is the ability of a material to resist impact loads as shown in Figure 9.3. Here, the toughness of a piece of high carbon steel in the soft (annealed) condition is compared with a piece of the same steel after it has been hardened by raising it to red heat and cooling it quickly (quenching it in cold water). The hardened steel shows a greater strength, but it lacks toughness. We will return to this important point later in this chapter.

A test for toughness, called the Izod test, uses a notched specimen that is hit by a heavy pendulum. The test conditions are carefully controlled, and the energy absorbed in bending or breaking the specimen is a measure of the toughness of the material from which it was made.

Elasticity

Materials that change shape when subjected to an applied force but spring back to their original size and shape when that force is removed, are said to be elastic. They have the property of elasticity.

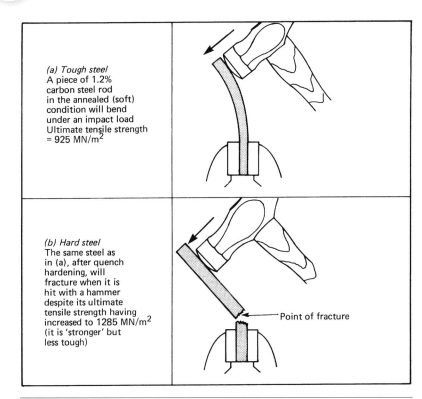

Figure 9.3 Impact loads.

Plasticity

Materials that flow into a new shape when subjected to an applied force and keep that shape when the applied force is removed are said to be plastic. They have the property of plasticity.

Ductility

Materials that can change shape by plastic flow when they are subjected to a pulling (tensile) force are said to be ductile. They have the property of ductility. This is shown in Figure 9.4a.

Malleability

Materials that can change shape by plastic flow when they are subjected to a squeezing (compressive) force are said to be malleable. They have the property of malleability. This is shown in Figure 9.4b.

Hardness

Materials that can withstand scratching or indentation by an even harder object are said to be hard. They have the property of hardness. Figure 9.5 shows the effect of pressing a hard steel ball into two pieces of metal with the same force. The ball sinks further into the softer of the two pieces of metal than it does into the harder.

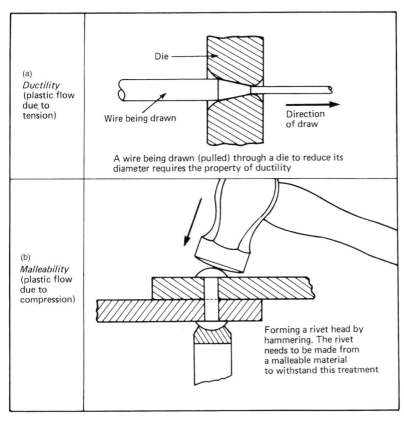

Figure 9.4 Two common engineering processes: drawing and riveting. Drawing exploits ductility while riveting exploits malleability.

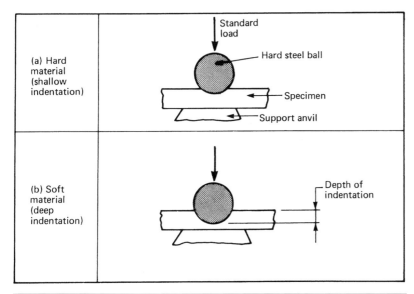

Figure 9.5 The effect of pressing a hard steel ball into two materials with different hardness properties.

There are various hardness tests available. The *Brinell hardness test* uses the principles set out above. A hardened steel ball is pressed into the specimen by a controlled load. The diameter of the indentation is measured using a special microscope. The hardness number is obtained from the measured diameter by use of conversion tables.

The *Vickers hardness test* is similar but uses a diamond pyramid instead of a hard steel ball. This enables harder materials to be tested. The diamond pyramid leaves a square indentation and the diagonal distance across the square is measured. Again, conversion tables are used to obtain the hardness number from the measured distance.

The *Rockwell hardness test* uses a diamond cone. A minor load is applied and a small indentation is made. A major load is then added and the indentation increases in depth. This increase in depth of the indentation is directly converted into the hardness number and it can be read from a dial on the machine.

Rigidity

Materials that resist changing shape under load are said to be rigid. They have the property of rigidity. The opposite of rigidity is flexibility. Rigid materials are usually less strong than flexible materials. For example, cast iron is more rigid than steel but steel is the stronger and tougher. However, the rigidity of cast iron makes it a useful material for machine frames and beds. If such components were made from a more flexible material the machine would lack accuracy and it would be deflected by the cutting forces.

Test your knowledge 9.4

Explain the difference between ductility and plasticity.

Test your knowledge 9.5

Explain the difference between soft and hard magnetic materials.

Learning outcome 9.3

State how the properties are affected by the application of heat

Thermal properties are to do with how a material responds to heat and different temperatures.

Thermal properties

Melting temperature

The melting temperature of a material is the temperature at which a material loses its solid properties. Most plastic materials and all metals become soft and eventually melt. Note that some plastics do not soften when heated, they only become charred and are destroyed.

Thermal conductivity

This is the ease with which materials conduct heat. Metals are good conductors of heat. Non-metals are poor conductors of heat. Therefore, non-metals are heat insulators. This will be considered later in Chapter 11.

Thermal expansion

Metals expand appreciably when heated and contract again when cooled. They have high coefficients of linear expansion. Non-metals expand to a lesser extent when heated. They have low coefficients of linear expansion. Again, these thermal properties will be considered in more detail in Chapter 11.

Activity 9.3

A gas turbine is to be fitted with a part manufactured from a material that has a melting point greater than 1250°C and a tensile strength greater than 700MPa. Use the matSdata database in order to select at least two suitable materials.

Learning outcome 9.4

Identify why the different properties make materials suitable for different applications

In this section we shall consider a number of common engineering materials and the properties that make them suitable for use in a range of engineering applications. The metals widely used in engineering are classified as shown in Figure 9.6 and we will start this section by looking at the ferrous metals that we first met in Chapter 8.

Ferrous metals

As previously stated, ferrous metals are based upon the metal iron. For engineering purposes iron is usually associated with

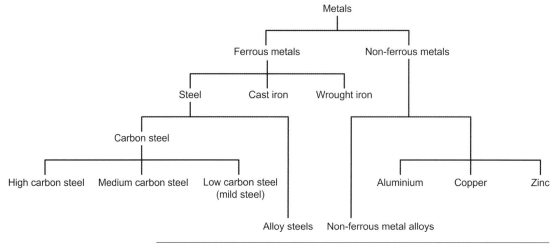

Figure 9.6 Classification of metals.

various amounts of the non-metal carbon. When the amount of carbon present is less than 1.8% we call the material steel. The figure of 1.8% is the theoretical maximum. In practice there is no advantage in increasing the amount of carbon present above 1.4%. We are only going to consider the plain carbon steels. Alloy steels are beyond the scope of this book. The effects of the carbon content on the properties of plain carbon steels are shown in Figure 9.7.

Cast irons are also ferrous metals. They have substantially more carbon than the plain carbon steels. Grey cast irons usually have a carbon content between 3.2 and 3.5%. Not all this carbon can be taken up by the iron and some is left over as flakes of graphite between the crystals of metal. It is these flakes of graphite that gives cast iron its particular properties and makes it a 'dirty' metal to machine.

Low carbon steels

Low carbon steels (also known as mild steels) are the cheapest and most widely used group of steels. Although they are the weakest of the steels, nevertheless they are stronger than most of the non-ferrous metals and alloys. They can be hot and cold worked and machined with ease. Low carbon steels are used for pressing out panels for car bodies and for general sheet-metal applications.

Medium carbon steels

These are harder, tougher, stronger and more expensive than the low carbon steels. They are less ductile than the low carbon steels and cannot be bent or formed to any great extent in the cold condition without risk of cracking. Greater force is required to bend

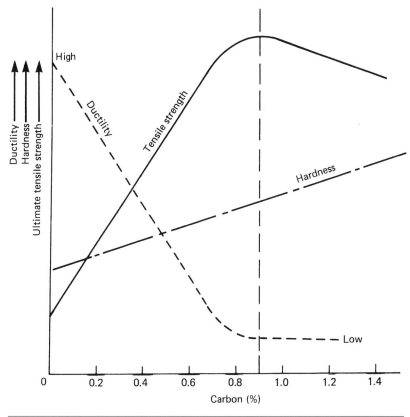

Figure 9.7 Effects of the carbon content on the properties of plain carbon steels.

and form them. Medium carbon steels hot forge well but close temperature control is essential. Low carbon steels (below 0.3%) can only be toughened by heating and quenching (cooling quickly by dipping in water). They cannot be hardened. Higher range carbon steels (above 0.6%) can be hardened and tempered by heating and quenching. Medium carbon steels (0.3% to 0.6%) are used in crank shafts, forgings, axles and in many other mechanically stressed components.

High carbon steels

These are harder, stronger and more expensive than medium carbon steels. They are also less tough. High carbon steels are available as hot rolled bars and forgings. Cold drawn high carbon steel wire (piano wire) is available in a limited range of sizes. Centre-less ground high carbon steel rods (silver steel) are available in a wide range of diameters (inch and metric sizes) in lengths of 333mm, 1m and 2m. High carbon steels can only be bent cold to a limited extent before cracking. High carbon steels are used in the manufacture of chisels, cutting tools and drill bits.

Key point

The term 'carbon steel' can be a little misleading because all steels contain carbon. Iron, the basic ingredient of steel, has so much carbon in its basic 'pig iron' form that carbon actually has to be removed to produce what we call 'carbon steel'. The strength and hardness of steel increases with the amount of carbon present.

Figure 9.8 A bicycle sprocket made from medium carbon steel.

Figure 9.9 High-quality adjustable spanners made from high carbon steel.

Test your knowledge 9.6

What type of steel would be required for the manufacture of each of the following:

a) a centre punch
b) a car body panel
c) a lathe bed
d) a woodworking chisel?

Non-ferrous metals

As you've seen, non-ferrous metals (i.e. metals that are not based on iron) include metals such as aluminium and zinc as well as alloys such as brass and bronze. We shall start by looking at copper, a material that is widely used in electrical and electronic engineering.

Copper

Pure copper is widely used for electrical conductors and switchgear components. It is second only to silver in conductivity but it is much more plentiful and very much less costly. Pure copper is too soft and ductile for most mechanical applications.

There are many other grades of copper for special applications. Copper is also the basis of many important alloys such as brass and bronze, and we will be considering these next. The general properties of copper are:

- relatively high strength
- very ductile so that it is usually cold worked – an annealed (softened) copper wire can be stretched to nearly twice its length before it snaps
- corrosion resistant

- second only to silver as a conductor of heat and electricity
- easily joined by soldering and brazing.

Copper is available as cold-drawn rods, wires and tubes. It is also available as cold-rolled sheet, strip and plate. Hot-worked copper is available as *extruded sections* and *hot stampings*. It can also be cast. Copper powders are used for making sintered components. It is one of the few pure metals of use to the engineer as a structural material.

Brass

Brass is an alloy of copper and zinc. The properties of a brass alloy and the applications for which you can use it depends upon the amount of zinc present. Most brasses are attacked by sea water. The salt water eats away the zinc (known as *dezincification*) and leaves a weak, porous, spongy mass of copper. To prevent this happening, a small amount of tin is added to the alloy.

Brass is a difficult metal to cast and brass castings tend to be coarse-grained and porous. Brass depends upon hot rolling from cast ingots, followed by cold rolling or drawing to give it its mechanical strength. It can also be hot extruded and plumbing fittings are made by hot stamping. Brass machines to a better finish than copper as it is more rigid and less ductile than that metal. You will find the different types of brass and their properties listed in the matSdata database.

Tin bronze

As the name implies, the tin bronzes are alloys of copper and tin. These alloys also have to have a deoxidizing element present to prevent the tin from oxidizing during casting and hot working. If the tin oxidizes the metal becomes hard and 'scratchy' and is weakened. The two deoxidizing elements commonly used are:

- zinc in gun-metal alloys.
- phosphorus in the phosphor-bronze alloys.

Unlike the brass alloys, the bronze alloys are usually used as *castings*. However, low tin content phosphor-bronze alloys can be extensively cold worked. Tin-bronze alloys are extremely resistant to corrosion and wear and are used for high-pressure valve bodies and heavy-duty bearings.

Aluminium

Aluminium has a density approximately one-third that of steel. However, it is also very much weaker so its strength/weight ratio is inferior. For stressed components, such as those found in aircraft, aluminium alloys have to be used. These can be as strong as steel and nearly as light as pure aluminium.

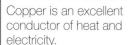

Key point

Copper is an excellent conductor of heat and electricity.

Key point

Aluminium (and its alloys) are lighter and less dense than iron and steel.

Key point

Pure aluminium is weak and soft but its mechanical strength can be increased by alloying with other metals. Such alloys have strengths that are comparable to steel.

High-purity aluminium is second only to copper as a conductor of heat and electricity. Aluminium is difficult to join by welding or soldering therefore aluminium electrical conductors are often terminated by *crimping*. Despite these difficulties, it is increasingly used for electrical conductors where its light weight and low cost compared with copper is an advantage. Pure aluminium is resistant to normal atmospheric corrosion but it is unsuitable for marine environments. It is available as wire, rod, cold-rolled sheet and extruded sections for heat sinks.

Commercially pure aluminium is not as pure as high-purity aluminium and it also contains up to 1% silicon to improve its strength and stiffness. As a result, it is not such a good conductor of electricity nor is it so corrosion resistant. It is available as wire, rod, cold-rolled sheet and extruded sections. It is also available as castings and forgings. Being stiffer than high-purity aluminium it machines better with less tendency to tear. It forms non-toxic oxides on its surface which makes it suitable for food processing plant and utensils. It is also used for forged and die-cast small machine parts.

Test your knowledge 9.7

Name three non-ferrous metals used in engineering and describe an application of each.

Learning outcome 9.5

Identify the basic heat treatment process as applied to changing the properties of materials

Earlier in Section 9.2 we mentioned that carbon steel can be made hard by raising it to red heat and then quenching it in cold water. The hardened steel shows a greater strength, but it lacks toughness. The properties of many metals and alloys can be changed by heating them to specified temperatures and cooling them under controlled conditions at specified rates. These are called, respectively, *critical temperatures* and *critical cooling rates*. Heat treatment is a complex subject so we are only going to consider the heat treatment of plain carbon steels in this book.

Hardening

The degree of hardness that can be given to any plain carbon steel depends upon two factors: the amount of carbon present, and how quickly the steel is cooled from the hardening temperature. The hardening temperature for medium carbon steels containing up to 0.8% carbon is bright red heat. Above 0.8% carbon the hardening temperature is dull red (cherry red) heat. Table 9.1a relates hardness to carbon content. Table 9.1b relates hardness to rate of cooling. Note that a number of safety rules must be observed when oil (rather than water or brine) is used as the quenching fluid.

Table 9.1 Heat treatment of steel.

a) Effect of carbon content

Type of steel	Carbon content (%)	Effect of quench hardening
Low carbon	< 0.3	Negligible
Medium carbon	0.3–0.5 0.5–0.9	Becomes tougher Becomes hard
High carbon	0.9–1.3	Becomes very hard

b) Rate of cooling

Carbon content (%)	Quenching media	Required treatment
0.3–0.5	Oil	Toughening
0.5–0.9	Oil	Toughening
0.5–0.9	Water	Hardening
0.9–1.2	Oil	Hardening

During the heat treatment process, a number of problems may occur. They are as follows:

Underheating

If the temperature of a steel does not reach its critical temperature, the steel won't harden.

Overheating

It is a common mistake to think that increasing the temperature from which the steel is quenched will increase its hardness. Once the correct temperature has been reached, the hardness will depend only upon the carbon content of the steel and its rate of cooling. If the temperature of a steel exceeds its critical temperature, grain growth will occur and the steel will be weakened. Also overheating will slow the cooling rate and will actually reduce the hardness of the steel.

Cracking

There are many causes of hardening cracks. Some of the more important are: sharp corners, sudden changes of section, screw threads, holes too near the edge of a component. These should all be avoided at the design stage as should over-rapid cooling for the type of steel being used.

Distortion

There are many causes of distortion. Some of the more important are as follows:

* lack of balance or symmetry in the shape of the component
* lack of uniform cooling – long thin components should be dipped end-on into the quenching bath
* change in the grain structure of the steel causing shrinkage.

No matter how much care is taken when quench hardening, some distortion (movement) will occur. Also slight changes in the chemical composition may occur at the surface of the metal. Therefore, precision components should be finish ground after hardening. The components must be left slightly oversize before grinding to allow for this. That is, a grinding allowance must be left on such components before hardening.

Tempering

Quench-hardened plain carbon steels are very brittle and unsuitable for immediate use. Therefore, further heat treatment is required. This is called tempering. It greatly increases the toughness of the hardened steel at the expense of some loss of hardness.

Tempering consists of reheating the hardened steel and again quenching it in oil or water. Typical tempering temperatures for various applications are summarized in Table 9.2.

You can judge the tempering temperature by the colour of the oxide film. First, the component must be polished after hardening and before tempering. Then heat the component gently and watch for the colour of the metal surface to change. When you see the appropriate colour appear, the component must be quenched immediately.

Key point

The degree of hardness that can be given to any plain carbon steel depends upon two factors: the amount of carbon present, and how quickly the steel is cooled from the hardening temperature.

Key point

Tempering temperature can be judged by the colour of the oxide film on the surface of the part or component undergoing heat treatment.

Table 9.2 Tempering temperatures.

Component	Temper colour	Temperature (°C)
Edge tools	Pale straw	220
Turning tools	Medium straw	230
Twist drills	Dark straw	240
Taps	Brown	250
Press tools	Brown/purple	260
Cold chisels	Purple	280
Springs	Blue	300
Toughening (medium carbon steels)	—	450–600

Test your knowledge 9.8

Explain the reason for tempering a quench-hardened plain carbon steel component.

Test your knowledge 9.9

Describe how you should harden and temper a small cold chisel made from 1.2% high carbon steel.

Review questions

1. Distinguish between the terms *ductility* and *malleability* when applied to engineering materials.

2. Explain what is meant by 'yielding' when applied to a material under stress.

3. If the length of a copper conductor is increased (without changing its cross-sectional area), what effect will this have on its electrical resistance?

4. With rubber parts and components, explain why it is usually necessary to avoid prolonged exposure to contact with oil.

5. Explain why some rubber and plastic materials degrade when exposed to sunlight.

6. What **two** materials can be added to pure iron in order to improve its magnetic properties?

7. On what **two** factors does the degree of hardness that can be given to any plain carbon steel depend?

8. Give examples of suitable non-ferrous metals (with reasons for your choice) for each of the following applications:
   a) an instrument case used to house a portable electronic test set
   b) a water pump to be used with a marine engine
   c) a 'bus-bar' to carry electric current in a steelworks
   d) a screw terminal used in an electric light fitting.

9. After heat treatment, explain why it is often necessary to finish hardened parts by grinding them to their final size.

10. Describe **two** different tests that can be applied to determine the hardness of a metal.

Chapter checklist

Learning outcome	Page number
9.1 State the properties associated with basic engineering materials.	226
9.2 Recognize the physical properties of engineering materials.	230
9.3 State how the properties are affected by the application of heat.	234
9.4 Identify why the different properties make materials suitable for different applications.	235
9.5 Identify the basic heat treatment process as applied to changing the properties of materials.	240

Numeracy

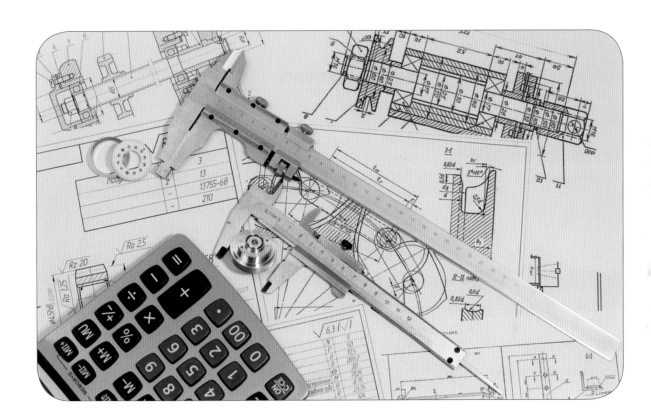

10.5   Calculate ratio, proportion and percentages.

10.6   Calculate area, surface area, mass, volume, capacity.

10.7   Calculate probability.

10.8   Calculate the square and square root of a number.

10.9   Transpose simple formulae.

10.10   Calculate speeds and feeds.

10.11   Construct simple graphs.

10.12   Calculate values for similar triangles.

10.13   Identify the techniques used for calculating approximation.

Chapter summary

Mathematics is a tool that engineers use to solve problems. Being numerate and able to use mathematics is therefore an essential skill that you must develop. This unit will provide you with sufficient mathematical knowledge for you to solve real engineering problems as well as a firm foundation that will enable you to make good progress in the other units. Once again, we've adopted a 'topic-based' approach in which the unit has been divided into sections, with each matched to one of the learning outcomes. The topics are introduced in a logical sequence and each section builds on those that have gone before. The sections deal with topics such as numbers, notation, fractions and decimals, units, statistics (average, mean, median and mode), ratio, proportion and percentage, mensuration (length, angle, area and volume), probability, squares and square roots, transposition of formulae, feeds and speeds, triangles, estimation and approximation. Worked examples and plenty of 'Test your knowledge' questions have been included to give you plenty of practice using mathematics to solve typical engineering problems. Finally, it's important that you have access to an electronic calculator throughout your study of this unit. Further details can be found in Appendix 2.

Learning outcome 10.1

Add, subtract, multiply and divide: whole numbers, fractions and decimals

Numbers

Let's begin with numbers. Engineers tend to use a lot of numbers for the simple reason that they *are* precise. For example, instead of saying that we need 'a large storage tank' we would convey a lot more meaning (and be much more *specific*) if we said that we need 'a tank with a capacity of 4500 litres'. In fact, when engineers draw up a *specification* they do so using numbers and drawings in preference to written descriptions.

Integers

Whole numbers that have a positive (+) sign, such as 1, 2, 3, 4..., are known as *positive integers*. Negative numbers (which have a sign), such as −1, −2, −3, −4..., are known as *negative integers*. Note that, if a number is positive, we don't usually include a positive (+) sign to show that it's positive. Instead, we simply assume that it's there!

The number of units that a number is from *zero* (regardless of its direction or sign) is known as its *absolute value*. Positive numbers are conventionally shown to the right of the *number line* (see Figure 10.1) while negative numbers are shown to the left. When the sign is shown these numbers are said to be *signed.* For example, the number 5 has an absolute value of 5 and its sign is positive. Similarly, the number −3 has an absolute value of 3 and its sign is negative. Note that the number zero (0) is unique in that it is neither a positive integer nor is it a negative integer.

Figure 10.1 The number line (showing positive and negative integers).

When performing arithmetic with negative numbers you need to be careful to show the signs. For example, if we need to find the sum of the first three positive integers (1, 2 and 3) we would write this as follows:

Sum of first three positive integers: 1 + 2 + 3 = 6

However, if we are asked to find the sum of the first three negative integers, 1, 2 and 3, we would write:

Sum of first three negative integers: (−1) + (−2) + (−3) = −1 −2 −3 = −6

Notice how we have used brackets to help to clarify the arithmetic. Brackets can be very useful as we shall see a little later on.

Test your knowledge 10.1

1 Find the sum of the first ten positive integers.
2 Find the sum of the series of negative integers from −5 to −10.

Notation

Notation is used in mathematics to simplify the writing of mathematical expressions and formulae. This notation is based on the use of symbols that you will already recognize including: = (equal), + (addition), − (subtraction), × (multiplication), and / or ÷ (division). Other symbols that you may not be so familiar with include ± (plus or minus), < (less than), > (greater than), ∝ (proportional to), ≈ (approximately equal), and √ (square root). You need to understand what each of these symbols means and how each of them is used.

Table 10.1 Symbols used in mathematical notation.

Symbol	Meaning
=	Equality
+	Addition
−	Subtraction
×	Multiplication
/ or ÷	Division
±	Plus or minus
<	Less than
>	Greater than
∝	Proportionality
≈	Approximately equal
√	Square root

Key point

When a positive number is divided by a negative number (or vice versa) the result is negative but when the two numbers are both negative the result is positive.

Key point

When a positive number is multiplied by a negative number the result is negative but when two negative numbers are multiplied together the result is positive.

Test your knowledge 10.2

Find the value of each of the following expressions:
a) $2 - 13$
b) $-7 + 19$
c) $5 \times (-9)$
d) $(-11) \times (-7)$
e) $15 \div (-3)$

Test your knowledge 10.3

Which of the following statements are true?
a) $1 + (-1) = 0$

b) $1 - (-1) = 0$
c) $-1 + (+1) = 0$
d) $-1 \times (+1) = 1$
e) $-1 \times (-1) = 1$
f) $1 \div (-1) = -1$
g) $-1 \div (-1) = 1$

Learning outcome 10.2

Convert fractions to decimals and decimals to fractions

Integers (or *whole numbers*) are often not precise enough for use in engineering applications. We can get over this problem in two ways; using *fractions* and using a *decimal point*. For example, the number that sits mid-way between the positive integers 3 and 4 can be expressed as a fraction, 3½, or by using a decimal point, 3.5. Similarly, the number that sits equally between −1 and −2 can be expressed as −1½ or −1.5 (see Figure 10.2). A table of some common fractions and their corresponding decimal values is shown in Table 10.2.

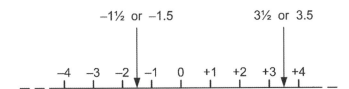

Figure 10.2 Decimal and fractional numbers shown on the number line..

Table 10.2 Fractions and their decimal values.

Fraction	Decimal value
½	0.5
¼	0.25
⅛	0.125
1/10	0.1
1/16	0.0625
1/100	0.01
1/1000	0.001

Test your knowledge 10.4

1 Express a) 0.75, b) 2.25 and c) 7.375 as fractional values.

2 Express a) 3¼, b) $\frac{5}{8}$ and c) $4\frac{3}{16}$ as decimal values.

3 Add 5¼ to 2½ and express your answer as a decimal.

4 Subtract 15¾ from 20½ and express your answer as a fraction.

5 Multiply 3.4 by 1.25 and express your answer as a decimal.

6 Multiply 4½ by ¾ and express your answer as a fraction.

7 Divide 6.25 by 0.5 and express your answer as a decimal.

8 What length of cable will remain when four 3¾m lengths of cable are cut from a 25m reel?

9 A thin steel plate has a thickness of 1/8 in. If 21 of these plates are stacked together, what thickness will the stack have? Express your answer as a decimal value.

10 A voltage of 0.015V is amplified by a factor 25. What will the resulting voltage be? Express your answer as a fraction of a volt.

Learning outcome 10.3

Identify the metric and imperial systems and the preferred standard form

Engineers need to be familiar with the ways of expressing units based on two different systems of measurement. Imperial units of measurement were commonly used in the UK up until about 50 years ago but since then, for most purposes, metric units are preferred. Imperial (and other units) are used in many other countries outside Europe such as the United States of America. As a result, you need to be able to convert metric to imperial units, and vice versa. Some metric and imperial units that you may already be familiar with are listed in Table 10.3.

In Table 10.3 you should notice that there are several different imperial units that can be used for a particular measurement. Just in case you are wondering, the relationship between these units is as follows:

1 lb = 16 oz

1 yd = 3 ft

1 ft = 12 in

1 gal = 8 pt.

Table 10.3 Familiar metric and imperial units.

Measurement	Metric unit	Imperial units
Length/distance	metre (m)	inch (in) foot (ft) yard (yd)
Mass/weight	gram (g)	ounce (oz) pound (lb) stone (st)
Volume/capacity	litre (l)	pint (pt) gallon (gal)

Multiples and sub-multiples

The metric system is much more convenient than the imperial system simply because it's based on units that can be easily multiplied or divided by 10, 100, 1000 and so on. This helps keep the arithmetic simple. So, for example, when dealing with length we have kilometres (thousands of metres) and millimetres (thousandths of metres). This is much easier than having to convert between yards, feet and inches.

Another important point is that the numbers that we meet in engineering can sometimes be very large or very small. For example, the surface coating on a metal component could be as thin as 0.000005m. At the other extreme, the weight of a bridge support could be as much as 5,000,000g. Needing to take into account all of these zeros can be a bit of a problem but we can make life a lot easier by using a standard range of multiples and sub-multiples. These use a *prefix* letter that adds a *multiplier* to the quoted value, as shown in Table 10.4.

Table 10.4 Some common multiples and sub-multiples.

Prefix	Abbreviation	Equivalent
giga	G	1,000,000,000
mega	M	1,000,000
kilo	k	1000
(none)	(none)	1
centi	c	0.01
milli	m	0.001
micro	μ	0.000,001
nano	n	0.000,000,001

Key point

To multiply a number by 1000 (one thousand) we move the decimal point three places to the right. To divide a number by 1000 we move the decimal point three places to the left.

Key point

To multiply a number by 1,000,000 (one million) we move the decimal point six places to the right. To divide a number by 1,000,000 we move the decimal point six places to the left.

It's worth noting that multiplying by 1000 is equivalent to moving the decimal point *three* places to the *right*. Dividing by 1000, on the other hand, is equivalent to moving the decimal point *three* places to the *left*. Similarly, multiplying by 1,000,000 is equivalent to moving the decimal point *six* places to the *right* while dividing by 1,000,000 is equivalent to moving the decimal point *six* places to the *left*.

Example 10.1

A high-speed vehicle test track has a total distance of 3.75km. Express this in metres.

To convert from km to m we need to multiply by 1000. Thus to convert 3.75km to m we multiply 3.75 by 1000, as follows:

$$3.75\text{km} = 3.75 \times 1000 = 3750\text{m}$$

(Note that we've effectively moved the decimal point three places to the right and added an extra zero on the end.)

Example 10.2

The current controlled by a transistor is 6.95mA. Express this in A.

To convert from mA to A we need to multiply by 0.001 (this is actually the same as dividing by 1000). Thus to convert 6.95mA to A we multiply 6.95 by 0.001, as follows:

$$6.95\text{mA} = 6.95 \times 0.001 = 0.00695\text{A}$$

(Note that we've effectively moved the decimal point three places to the left and added some leading zeros.)

Test your knowledge 10.5

Express:

1 752mm in cm
2 175,250Hz in kHz
3 4390ms in s
4 0.00047V in μV

Conversion of metric and imperial units

Converting between the different units of measurement is usually quite straightforward. You just need to know the appropriate

Table 10.5 Some useful conversion factors.

Measurement	To convert...	Multiply by...
Length/distance	Metres (mm) to feet (ft) Millimetres (mm) to inches (in)	3.2808 0.0394
	Feet (ft) to metres (m) Inches (n) to millimetres (mm)	0.3048 25.4
Area	Square metres (m²) to square feet (ft²) Square millimetres (mm²) to square inches (in²)	10.7635 0.00155
	Square feet (ft²) to square metres (m²) Square inches (in²) to square metres (m²)	0.0929 645.16
Volume/capacity	Litres (l) to UK pints (pt) Litres (l) to UK gallons (gal)	1.76 0.22
	UK pints (pt) to litres (l) UK gallons (gal) to litres (l)	0.5683 4.546
Mass/weight	Grams (g) to ounces (oz) Kilograms (kg) to pounds (lb)	0.0353 2.2046
	Ounces (oz) to grams (g) Pounds (lb) to kilograms (kg)	28.35 0.4536

conversion factor. If the arithmetic gets difficult because the conversion factor isn't a whole number you can always use a calculator. Appendix 2 on page 336 shows you how to use one. Table 10.5 shows some useful conversion factors. Appendix 3 on page 339 will help you with fractional and decimal conversion of metric and imperial units.

Test your knowledge 10.6

1 A beam has a length of 2.5 metres. What is its length in feet?
2 A cable has a length of 42 feet. Express this in metres.
3 A tarpaulin has an area of 12 square metres. What is its area in square feet?
4 A lorry has a fuel tank with a capacity of 45 UK gallons. How many litres of fuel would be required to fill the tank?
5 A single steel plate weighs 12kg. If 24 similar plates are to be loaded onto a fork lift truck what is the total load expressed in pounds?

> **Learning outcome 10.4**
>
> **Calculate average, mean, median and mode**

When we have to deal with a series of numbers, we might need to know what the average value of those numbers is, or we might need to know which value occurs most often, or which value lies in the middle of the range. A typical situation in which this might occur is when checking to see how closely a number of manufactured parts conform to their specification.

Mean or average

The mean is simply the average of several numbers. You can easily find the mean of several numbers by simply adding them up before dividing the total by how many numbers there are. So, for example, the average of the numbers 5, 1, 6, 11 and 7 is 6. This is calculated as follows:

$$\text{Mean or average value} = \frac{\text{sum of values}}{\text{number of values}}$$

$$= \frac{5+1+6+11+7}{5} = \frac{30}{5} = 6$$

Median

The median is the middle value in a sorted list of numbers. You can find the median by sorting the numbers into ascending order and then locating the middle value. If you have five numbers in the list (as in the previous example) the middle value would be the third in the list. So, to find the median value of 5, 1, 6, 11 and 7 we would first rearrange the list into ascending order to arrive at 1, 5, 6, 7 and 11. The middle of the list (the third number) is 6 so the median value is 6 (and in this case it is the same as the average value). But this isn't always the case!

If there's an odd number of values in a sorted list it's easy to find the middle value. If there's an even number you can calculate the median by taking the two values either side of where the middle value would be and finding their mean value.

Here's an example of how this works. To find the median of the following six values 3, 7, 8, 2, 4 and 7 we would first sort this list, as before, to arrive at 2, 3, 4, 7, 7 and 8. Now we add the two middle values in the sorted list (4 and 7) and their average value is (4 + 7)/2 = 11/2 = 5½. Thus the median of this new list is 5½.

Mode

The mode in a set of numbers is simply the value that occurs most often in the list. For example, the mode of 5, 6, 4, 3, 4, 7, 4, 3 is 4 because this value occurs more than any other. When determining the mode of a large set of numbers it can be useful to first arrange the values in ascending order before counting the number of occurrences of each value.

Here's an example. Let's suppose that we need to find the mode of these values: 5, 7, 7, 8, 6, 9, 3, 5, 4, 8, 2, 4, 7, 6, 4, 7, 6 and 5. We can write these in ascending order and then count the occurrences of each:

2	Total: 1
3	Total: 1
4, 4, 4	Total: 3
5, 5, 5	Total: 3
6, 6, 6	Total: 3
7, 7, 7, 7	Total: 4 therefore the mode is 7
8, 8	Total: 2
9	Total: 1

Test your knowledge 10.7

1 Find the average of the following numbers: 3, 7, 8, 2, 4 and 7.

2 Find the mean and median of the following numbers: 10, 8, 8, 15, 11, 15, 9 and 12.

3 Find the mean and mode of the following numbers: 12, 15, 25, 18, 17, 11, 17, 10, 8 and 17.

Learning outcome 10.5

Calculate ratio, proportion and percentages

Engineers are sometimes more concerned with how things are shared or divided than what their actual values are. For example, we might need to cut a reel of cable into four equal lengths or we might need to know what proportion of manufactured parts fail to comply with a particular specification. Because of this you need to have an understanding of ratio, proportion and percentage.

Ratio

Ratio tells us how a whole number is divided into different parts. Note that we are often not concerned with the actual amount but more about the proportions into which it is divided. This may sound a whole lot more complex than it is, so here are a few examples to show you how it works:

- A two-stroke engine requires a fuel mixture of 40:1. This means that one part of two-stroke oil must be mixed with every 40 parts of petrol. The ratio should be the same for any quantity of fuel.
- A gearbox has a ratio of 4 to 1. This means that, for every four turns of the input shaft the output shaft will just make one turn. The ratio is not affected by engine speed. If the input speed increases so will the output speed.

Proportion

In engineering applications, when one quantity changes it normally affects a number of other quantities. For example, as long as we remain in the same gear ratio, when the engine speed of a car increases its road speed will also increase. To put this in a mathematical way, we can say that road speed is *directly proportional* to engine speed. Using mathematical notation and symbols to represent the quantities, we would write this as follows:

$$v \propto N$$

where v represents road speed and N represents the engine speed in revolutions per second.

In some cases, an increase in one quantity produces a *reduction* in another quantity. For example, if the road speed of a car increases, the time taken for it to travel a measured distance will decrease. To put this in a mathematical way we would say that time taken to travel a measured distance is *inversely proportional* to road speed. Once again, using mathematical notation and symbols to represent the quantities, we would write this as follows:

$$t \propto \frac{1}{v}$$

where t represents the time taken and v represents the road speed.

Once again this might sound a little difficult so here are a few examples to show you how it works:

- The mass of a steel I-beam is directly proportional to its length. If a 1m length of the beam has a mass of 15kg the mass of a 4m long beam will be $15 \times 4 = 60$kg.
- The resistance of a copper bus bar is inversely proportional to its cross-sectional area. If a fixed length of the bus bar has a

resistance of 0.1Ω and an area of 10mm^2 by doubling the area to 20mm^2 we would reduce the resistance by half to 0.05Ω.

Percentage

Per cent simply means 'per hundred'. So 10 per cent means 'ten parts in a hundred', 20 per cent means 'twenty parts in a hundred', and so on. Like ratios, percentages are used without reference to the actual value of the quantity concerned. Here are some typical examples of how we use percentages:

- A heat exchanger has an efficiency of 90%. This means that, for every 100 joules of input, only 90 joules will be usefully converted and the remaining 10 joules will be lost.
- A manufacturer of spare parts offers a 20% discount on all orders. This means that, if a purchaser places an order for £1000 worth of spare parts, the discount will amount to £200 and the purchaser only needs to pay £800.

Key point

If the value of one variable increases when the value of another variable increases we say that they are proportional to one another. If the value of one variable increases when the value of another variable decreases we say that they are inversely proportional to one another.

Test your knowledge 10.8

1 A gearbox has a ratio of 15:1. If the input shaft turns at a speed of 360 revolutions per minute (RPM) what will the output shaft speed be?

2 When a batch of 60 power supplies is tested, 15 are found to be faulty. What is the ratio of working units to faulty units?

3 The 2.4kNm torque produced by an engine is divided between two shafts. If one shaft receives 1.44kNm what torque will be supplied to the other shaft and by what ratio is the input torque divided?

4 Three out of every four bolts in a box are found to be corroded. If the box contains 120 bolts how many will be usable?

5 When a batch of 20 pilot lamps is tested, 15% are found to be faulty. How many lamps are working?

6 An electronic solder consists of a 65% tin and 35% lead by weight. How much lead is present in a 2kg reel of solder?

Test your knowledge 10.9

1 The density, ρ, of a body is directly proportional to its mass, m, and inversely proportional to its volume, V. Use the symbols, ρ, m and V, to write down an expression for density in terms of mass and volume.

2 A block of polyurethane foam has a mass of 0.155kg and a volume of 1.2m³. Determine the density of the alloy (in kg/m³) using the relationship that you obtain in Question 1. Express your answer correct to four decimal places.

Learning outcome 10.6

Calculate area, surface area, mass, volume, capacity

Being able to make measurements accurately is important in all branches of engineering. Mensuration is the name given to the science of measurement and in this section we will be looking at the ways in which we measure and quantify length, area, shape and volume.

Length, area and volume

Length, area and volume are extremely important in engineering and engineers frequently have to carry out calculations involving these quantities. Measurement of length is often carried out with a rule, but when we need more accurate measurements, particularly of small components, we often use a digital micrometer or caliper (see Figure 10.3).

Figure 10.3 A digital micrometer.

Area and volume

Area and volume can be calculated from measurements of length but, as we shall see later in this section, the way that we calculate area or volume depends on the particular shape that we are dealing with. Furthermore, recognizing how more complex shapes can be made from basic shapes is a particular skill that engineers need.

Shapes

Figure 10.4 shows several common shapes which have different numbers of straight sides. Notice that the triangle has three sides, the square has four sides, the pentagon has five sides, and so on. And, although the circle does not have straight lines, it can actually be considered to be an object with an infinite number of sides of equal length!

Angular and linear measure

Being able to measure angles as well as length is important in many engineering applications. The essential difference between angular measure and linear measure is illustrated in Figure 10.5.

Shape	Name	Number of sides
	Triangle	3
	Square	4
	Pentagon	5
	Hexagon	6
	Septagon	7
	Octagon	8
	Nonagon	9
	Circle	infinite

Figure 10.4 Some common shapes.

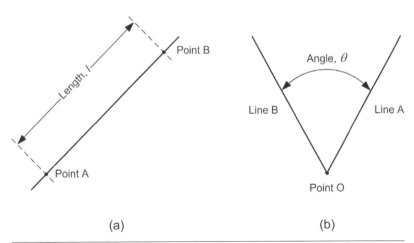

(a)   (b)

Figure 10.5 Linear and angular measure.

In Figure 10.5a we are concerned with the distance between points A and B measured along a straight line which joins the two points. In Figure 10.5b, we are concerned with the amount of rotation between lines A and B (which can be of any length) which meet at point O.

One complete rotation, starting at point X and returning to point X in Figure 10.6a, is equivalent to an angle of 360°. Some other angles are illustrated in Figures 10.6b to 10.6d. When lines are perpendicular to one another (i.e. at right angles) the angle between them is 90°, as shown in Figure 10.7.

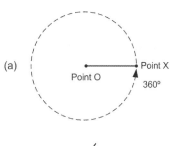

(a)

Point X
Point O
360°

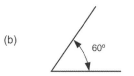

(b)

60°

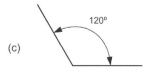

(c)

120°

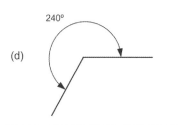

(d)

240°

Figure 10.6 One complete rotation is equivalent to moving through an angle of 360°.

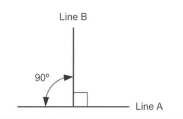

Line B

90°

Line A

Figure 10.7 Perpendicular lines.

Key point

A right angle is an angle of 90°.

Area

The formulae for determining areas of various shapes are shown in Figure 10.8. Some simple rectangular shapes are shown in Figure 10.9. In Figure 10.9a the shape is a square measuring 1m × 1m. The area of the shape is thus $1 \times 1 = 1m^2$. In Figure 10.9b the shape is a square measuring 2m × 2m. The area of the shape is thus $2 \times 2 = 4m^2$. In Figure 10.9c the shape is a square measuring 3m × 3m. The area of the shape is thus $3 \times 3 = 9m^2$.

In Figure 10.9d we are dealing with a rectangle (rather than a perfect square). The rectangle has dimensions 4m × 3m and its area is $4 \times 3 = 12m^2$. Finally, in Figure 10.9e we have a shape that can be divided into two rectangles. The dimensions of one rectangle is 2m × 3m while the other is a perfect square measuring 2m × 2m. The area of the shape is the sum of the areas of the two rectangles, i.e. $(2 \times 3) + (2 \times 2) = 6 + 4 = 10m^2$. Later in Section 10.12 we will be looking at triangles in more detail.

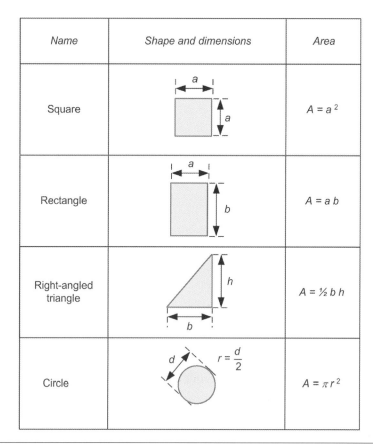

Name	Shape and dimensions	Area
Square	a ... a	$A = a^2$
Rectangle	a ... b	$A = ab$
Right-angled triangle	h ... b	$A = \frac{1}{2}bh$
Circle	d ... $r = \frac{d}{2}$	$A = \pi r^2$

Figure 10.8 Formulae for determining the area of various common shapes found in engineering.

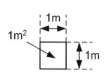

Total area = 1 x 1 = 1m²

(a)

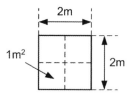

Total area = 2 x 2 = 4m²

(b)

Key point

Perpendicular lines are at 90° to one another.

Key point

In moving through one complete revolution a shaft moves through an angle of 360°.

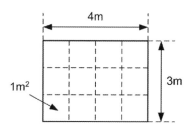

Total area = 3 x 3 = 9m²

(c)

Total area = 4 x 3 = 12m²

(d)

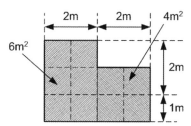

Total area = (2 x 3) + (2 x 2) = 6 + 4 = 10m²

(e)

Figure 10.9 Area of some rectangular shapes.

Test your knowledge 10.10

Find the area of each of the shapes shown in Figure 10.10.

Circles

The circle is another shape that is often found in engineering and you should be able to think of quite a few applications of this particular shape! The properties of a circle are illustrated in Figure 10.11 and summarized in Table 10.6. Note that π is a *constant* and for most purposes we can use $\pi = 22/7$ or 3.142.

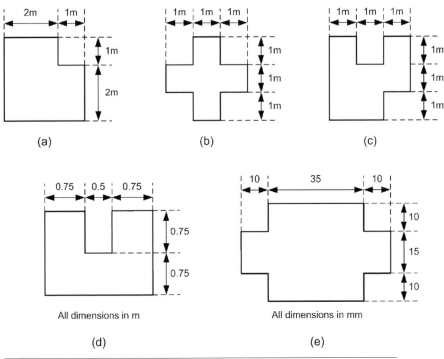

(a)   (b)   (c)

All dimensions in m   All dimensions in mm

(d)   (e)

Figure 10.10 See Test your knowledge 10.10.

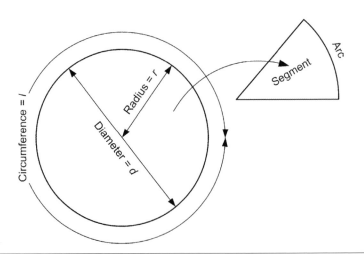

Figure 10.11 Properties of a circle.

Table 10.6 Properties of a circle.

Property	Relationship
Radius, r	$r = d/2$
Diameter, d	$d = 2r$
Circumference, l	$l = \pi d$
Area, A	$A = \pi r^2$

Example 10.3

A circular test track is to be constructed. If the test track is to have a circumference of 2.5km, determine the required diameter of the circle. The relationship between the circumference, l, of a circle and its diameter, d, is as follows:

$$l = \pi d$$

Rearranging this in order to make d the subject of the equation gives:

$$d = \frac{l}{\pi}$$

and since $l = 2500$m we have:

$$d = \frac{2500}{3.142} = 796\text{m}$$

Example 10.4

Find the surface area of the washer shown in Figure 10.12.

The area of the washer can be found by subtracting the area of a circle having a radius of 5mm from a circle that has a radius of 8mm.

Area of a circle having a radius of 8mm:

$$A_1 = \pi r^2 = \pi \times 8^2 = 3.142 \times 64 = 201.6\text{mm}^2$$

Area of a circle having a radius of 5mm:

$$A_2 = \pi r^2 = \pi \times 5^2 = 3.142 \times 25 = 78.54\text{mm}^2$$

Thus the area of the washer is given by:

$$A = A_1 - A_2 = 201.06 - 78.54 = 122.52\text{mm}^2$$

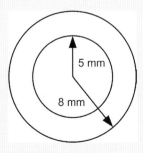

Figure 10.12 See Example 10.4.

Test your knowledge 10.11

1 A ceramic disc has a diameter of 19mm. Determine the surface area of the disc.
2 A spacecraft travels in a perfect circular orbit around a planet. If the spacecraft operates at a distance of 5500km from the centre of the planet, calculate the total distance travelled by the spacecraft when making one complete orbit of the planet.

Volume

Area is a two-dimensional property which is found my multiplying one length by another length while volume is a three-dimensional property found by multiplying area by a length. Figure 10.13 illustrates this important concept. The volumes of some common objects are listed in Table 10.7.

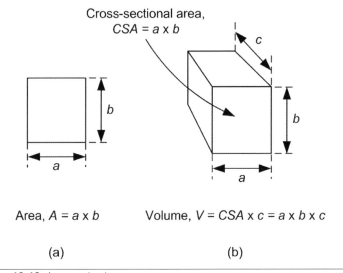

Cross-sectional area, $CSA = a \times b$

Area, $A = a \times b$ Volume, $V = CSA \times c = a \times b \times c$

(a)   (b)

Figure 10.13 Area and volume.

Table 10.7 Volumes of various solids.

Object	Formula
Cube with side length $= s$	$V = s^3$
Rectangular box or block with dimensions a, b, c	$V = abc$
Cylinder with height h and radius r	$V = \pi r^2 h$
Sphere with radius r	$V = \dfrac{4}{3}\pi r^3$

Example 10.5

Determine the volume of the trailer shown in Figure 10.14.

The volume of a rectangular box (see Table 10.7) is given by:

$$V = a\,b\,c$$

and from Figure 10.14:

$a = 2.4$m, $b = 6.8$m, and $c = 3.4$m

Hence $V = 2.4 \times 6.8 \times 3.4 = 55.5\text{m}^3$

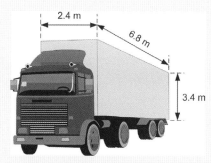

Figure 10.14 See Example 10.5.

Example 10.6

Determine the volume (in cubic centimetres) of a cylindrical block of alloy having a diameter of 24cm and a height of 40cm.

The volume of a cylinder (see Table 10.7) is given by:

$$V = \pi\, r^2\, h$$

Since the diameter, $d = 24$cm, the radius, r, will be given by:

$$r = \frac{r}{2} = \frac{24}{2} = 12\,\text{cm}$$

The volume (in cubic centimetres) can be calculated as follows:

$$V = \pi\, r^2\, h = 3.142 \times 12^2 \times 40 = 18{,}098\text{cm}^3$$

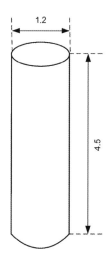

All dimensions in m

Figure 10.15 See Test your knowledge 10.12

Test your knowledge 10.12

Determine the volume of the concrete pillar shown in Figure 10.15.

Test your knowledge 10.13

Determine the internal volume of the equipment cabinet shown in Figure 10.16.

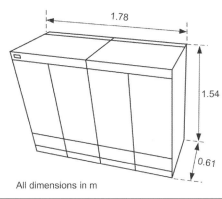

1.78
1.54
0.61
All dimensions in m

Figure 10.16 See Test your knowledge 10.13.

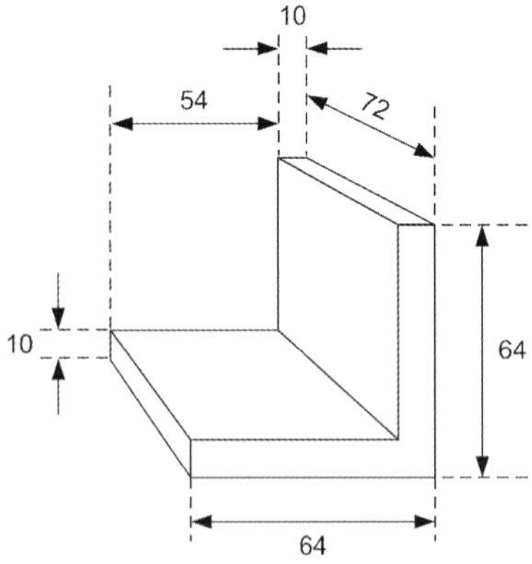

10
54
72
10
64
64
All dimensions in mm

Figure 10.17 See Test your knowledge 10.14.

Test your knowledge 10.14

Determine the volume of metal used to make the right-angled alloy flange shown in Figure 10.17.

Mass and weight

Mass is defined as the *quantity of matter* in a body. It's important to be aware that the mass of a body remains the same regardless of where the body is. So, for example, a mass of 50kg will be the same on the surface of the Earth as it will be in outer space (where there is *zero gravity*).

The weight of a body is determined by its mass and the gravitational force acting on the body. So, if there is no gravitational force (for example, in outer space) then a body will have no weight! However, in most practical cases we are concerned with what things weigh on the surface of the Earth in which case the relationship between mass and weight is given by:

$$W = mg$$

where W is the weight in newtons (N), m is the weight in kg, and g is the gravitational acceleration (in m/s^2). On the surface of the Earth, g is a constant equal to 9.81m/s^2. In Unit 11 we will be looking at this in more detail.

Key point

Mass and weight are not the same thing! The mass of an object remains the same wherever it is whereas the weight of a body is determined by the product of its mass and the gravitational force that is acting on it.

Test your knowledge 10.15

1. Calculate the weight of a machined part that has a mass of 3.5kg.
2. A platform is rated for a maximum loads of 2.5kN. How many identical blocks, each having a mass of 18kg, can be safely placed on it?

Learning outcome 10.7

Calculate probability

Probability is the measure of the likelihood that an event will occur. The probability of something happening can be expressed on a scale that ranges from 0 to 1. A probability of zero means that the event can never happen while a probability of 1 means that it is certain to happen. The probability of a coin toss returning a head on the next throw is 0.5 (or 50%). Similarly, the probability of it returning a tail on the next throw is also 0.5 (or 50%). In other words, it is just as likely that a head will result as it is for a tail to occur. Let's look at the maths behind this. We could express these outcomes in the form of a simple formula where P represents probability and the outcome is stated within brackets. So:

$$P_{(head)} = 0.5 \text{ and } P_{(tail)} = 0.5$$

Because a coin must land one way up or the other (discounting, the possibility that the coin could land on its edge!), the probability of *either* a head *or* a tail must be 1. In other words, there is certain to be an outcome but the chances of you getting a head or a tail prediction right is only 50%. In other words, after a very large number of coin throws your prediction of the outcome would be right 50% of the time.

The value of probability, P, is defined simply as:

$$P = \frac{\text{number of wanted outcomes}}{\text{number of possible outcomes}}$$

The faces of a die have values from 1 to 6. There are thus six possible outcomes. If you wanted to throw a six we could calculate the probability as follows:

$$P = \frac{\text{number of wanted outcomes}}{\text{number of possible outcomes}} = \frac{1}{6} \approx 0.167$$

Note that the number of wanted outcomes is 1 since we only want the die to display a six.

The probability of throwing a five *or* a six would be calculated as follows:

$$P = \frac{\text{number of wanted outcomes}}{\text{number of possible outcomes}} = \frac{2}{6} = \frac{1}{3} \approx 0.333$$

Now what would be the chances of throwing two sixes in succession? These two events would be independent because the outcome of the first event would not have any effect on the second.

$$P_{(two\ sixes)} = P_{(six)} \times P_{(six)} \approx 0.167 \times 0.167 = 0.027$$

Thus the chances of throwing two sixes in succession is less than 3%.

Let's take another example. In a standard pack of cards there are 52 cards with 13 cards in each suit (spades, hearts, clubs and diamonds). There are two suits of red cards (hearts and diamonds) and two suits of black cards (spades and clubs). Thus, when drawing cards from a full pack you would have 50% chance of drawing a red card and similarly you would have the same chance of drawing a black card. Thus $P_{(red)} = 0.5$ and $P_{(black)} = 0.5$.

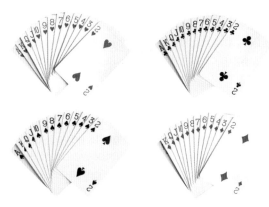

Figure 10.18 A pack of cards has four suits with 13 cards in each suit.

The probability of drawing any particular suit (spade, heart, club, diamond) would be 13/52 or 25%. Thus $P_{(spade)} = 0.25$, $P_{(heart)} = 0.25$, $P_{(club)} = 0.25$ and $P_{(diamond)} = 0.25$.

The probability of drawing two hearts in a row (but without replacing them in the pack) would be calculated as follows:

$$P_{(two\ hearts)} = P_{(heart)} \times P_{(heart)} = \frac{13}{52} \times \frac{12}{51} \approx 0.25 \times 0.235 \approx 0.059$$

Thus the probability of drawing two hearts in succession is less than 6%. Note how the probability of choosing a heart on the second draw is less than that on the first draw. The reason for this is that we've changed the odds by reducing the number of red cards (from 13 to 12) and the total number of cards (from 52 to 51). Because the number of cards has changed the two events are no longer independent.

Key point

The higher the probability of an event, the more certain we are that the event will occur. A simple example is the toss of a coin. In this case the two outcomes are equally probable because the probability of a head is the same as that of a tail.

Key point

The term 'probability' refers to the likelihood of an event occurring. Probability can be expressed as a percentage or a number between 0 and 1. A probability of zero means that the event can never happen while a probability of 1 means that it is certain to happen.

Test your knowledge 10.16

A box contains 24 lamps of which three are known to be faulty. If a lamp is taken randomly from the box, what is the probability of it working?

Learning outcome 10.8

Calculate the square and square root of a number

When a number is multiplied by itself we say that it is *squared*. Conversely, when a number is multiplied by itself to equal another number we say that the first number is the *square root* of the second number. Since we often use squares and square roots in engineering it's important to know how they work.

Squares

The number 4 is the same as 2×2, that is, 2 multiplied by itself. We can write (2×2) as 2^2. In words, we would call this 'two raised to the power two' or simply 'two squared'. Thus: $2 \times 2 = 2^2$. Similarly, $3 \times 3 = 3^2$, $4 \times 4 = 4^2$, $5 \times 5 = 5^2$, and so on. From this it follows that:

$$2^2 = 4, \ 3^2 = 9, \ 4^2 = 16 \text{ and } 5^2 = 25$$

Applying the rule for multiplying negative numbers that we met earlier in Section 10.1 it follows that the square of a negative number will take a positive value. Thus:

$(-2)^2 = (-2 \times -2) = 4$, $(-3)^2 = (-3 \times -3) = 9$, $(-4)^2 = (-4 \times -4) = 16$ and $(-5)^2 = (-5 \times -5) = 25$

Square roots

The square root of a number is that number which, when multiplied by itself will be equal to the original number. So, for example, because $2 \times 2 = 4$ the square root of 4 is 2. Similarly, the square root of 9 is 3, the square root of 16 is 4 and the square root of 25 is 5. We use the symbol $\sqrt{\ }$ to denote the square root so we can write:

$$\sqrt{4} = 2, \ \sqrt{9} = 3, \sqrt{16} = 4 \text{ and } \sqrt{25} = 5$$

There's just one problem with this. There are always two answers when finding the square root of a number. This is because when two negative numbers are multiplied together the result becomes positive. So, for example, if you multiply -2 by -2 the result is $(-2) \times (-2) = +4$. Similarly, $(-3) \times (-3) = +9$, and so on.

To take this into account we can add a 'plus or minus' sign to the result, like this:

$$\sqrt{4} = \pm 2, \ \sqrt{9} = \pm 3, \sqrt{16} = \pm 4 \text{ and } \sqrt{25} = \pm 5$$

Key point

The square of a number, regardless of whether it is positive or negative, will always be positive.

Key point

When finding the square root of a number there will always be two possible values. One of these is positive and the other is the same but negative.

Fortunately, most calculators have square, x^2, and square root, $\sqrt{x}$ buttons to make calculations easy.

Test your knowledge 10.17

1 Use a calculator to find the value of a) 4.4^2 and b) 27^2.

2 Use a calculator to find the value of a) $\sqrt{5.25}$ and b) $\sqrt{1976}$

3 Find the two values for x that satisfy the equation
$$x = 13.5^2 - \sqrt{7680}$$

Learning outcome 10.9

Transpose simple formulae

Being able to change the subject of an equation (or *formula*) is quite important. We call this *transposition* and it allows us to make the quantity that we are trying to find the subject of the equation. All we need to remember is that whatever we do to one side of an equation we must do to the other side, as shown in the next few examples.

Example 10.7

Make t the subject of the equation, $v = u + a\,t$

Let's start by subtracting u from both sides of the equation, thus:
$$v - u = u + a\,t - u$$

Regrouping the terms on the right-hand side (RHS) gives:
$$v - u = u - u + a\,t$$

But $u - u = 0$ so we can reduce the RHS to:
$$v - u = 0 + a\,t$$

So we arrive at:
$$v - u = a\,t$$

Next we must divide both sides by a:
$$\frac{v-u}{a} = \frac{at}{a}$$

Now $\dfrac{a}{a} = 1$ so:
$$\frac{v-u}{a} = t$$

Finally, exchanging the LHS and RHS gives:

$$t = \frac{v - u}{a}$$

Note that it's not necessary to show all of the intermediate stages when working through the transposition (as we have done in this first example).

Example 10.8

Make v the subject of the equation, $h = \dfrac{v^2}{2g}$

Start by multiplying both sides by $2g$, as follows:

$$h \times 2g = \frac{v^2}{2g} \times 2g$$

Now $\dfrac{2g}{2g} = 1$ so the RHS becomes just v^2, as follows:

$$h \times 2g = v^2$$

Rearranging this expression and exchanging the LHS and RHS gives:

$$v^2 = 2gh$$

Finally, taking the square root of both sides gives:

$$\sqrt{v^2} = \sqrt{2gh}$$

But since $\sqrt{v^2} = v$ we can arrive at the final version of our transposed equation:

$$v = \sqrt{2gh}$$

Test your knowledge 10.18

1 Make m the subject of the equation: $W = mg$

2 Make Z the subject of the equation: $I = \dfrac{V}{Z}$

3 Make C the subject of the equation: $X = \dfrac{1}{2\pi f C}$

4 Make u the subject of the equation: $s = ut + \dfrac{1}{2}at^2$

5 Make L the subject of the equation: $f = \dfrac{1}{2\pi\sqrt{LC}}$

Test your knowledge 10.19

1 A force of 50N is applied to an oil seal having an area of 0.0045m². Given that pressure is equal to force divided by area, determine the pressure acting on the seal. Express your answer in pascals (Pa).

2 A car travelling at 50m/s accelerates uniformly to reach a speed of 75m/s in a time of 6.8s. Given that $v = u + at$ (where v is the final velocity, u is the initial velocity, a is the acceleration and t is the time) make a the subject of the equation and then determine the acceleration. Express your answer in m/s².

Learning outcome 10.10

Calculate speeds and feeds

When using a machine tool the cutting process is affected by two important factors: the speed of cutting and rate at which the work is fed. Figure 10.19 shows a typical cutting process where the work rotates in the jaws of a chuck while the cutting tool moves over its surface.

Figure 10.19 Work being turned on a lathe.

Cutting speed

Cutting speed is the speed (or *velocity*) between the cutting tool and the surface of the piece of work (the *workpiece*) that it's cutting. When a lathe performs a *turning* operation the speed (or *angular velocity*) at which the workpiece rotates is important. This speed is called the *spindle speed*, *S*, and it is usually expressed in terms of the number of revolutions of the machine's spindle that occur in one minute (i.e. RPM or rev/min). We've illustrated all of this in Figure 10.20.

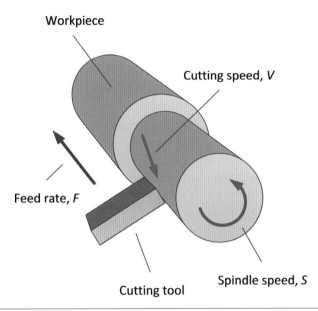

Figure 10.20 Cutting speed and surface speed.

Cutting speed is expressed in units of distance moved around the circumference of the workpiece surface per unit of time. In the imperial system it is often expressed in terms of feet per minute (ft/min) while in the metric system it is usually expressed in metres per minute (m/min). Note that the cutting force is applied at a tangent to the workpiece cross-section, as shown in Figure 10.20.

There is a linear relationship between the spindle speed and the cutting speed. The circumference of the workpiece can be found by multiplying its diameter, *D*, by the constant, π (3.142 approx.). If the diameter of the workpiece is measured in mm, in one complete revolution of the spindle the cutting distance will be ($\pi \times D$)mm and since there are 1000mm in 1m we can use the following relationship to calculate the cutting speed, *V*:

$$V = \frac{\pi DS}{1000} \text{ m/min}$$

where *S* = spindle speed (RPM) in rev/min and *D* = workpiece diameter in mm.

We can rearrange this formula (see earlier in Section 10.9) to make the spindle speed, S, the subject, as follows:

$$S = \frac{1000V}{\pi D} \text{ rev/min}$$

Feed rate

While the workpiece is rotating, the cutting tool moves along the surface of the workpiece from one end to the other so there's another speed that we need to be aware of. This is referred to as the *feed rate* and it is the speed (or *velocity*) at which the cutting tool is advanced along the workpiece.

The correct feed rate depends on the motion of the tool and workpiece; when the workpiece rotates (e.g. in turning and boring) the units are usually expressed in distance per spindle revolution, either inches per revolution using the imperial system or millimetres per revolution in the metric system. The feed rate, F, expressed in mm/min can be calculated from:

$$F = f \times S \text{ mm/min}$$

where f = feed rate in mm/rev and S = spindle speed (RPM) in rev/min.

Speeds and feeds for different materials

Cutting speeds and feed rates depend on the material that's being cut, the required finish, and the type, material and size of cutting tool that's used. In order to make life easy, tables and computer applications are invariably used to help select the correct cutting parameters. Table 10.8 gives a range of typical cutting speeds and feed rates for different materials when a plain high-speed steel

Table 10.8 Some typical cutting speeds and feed rates for various metals.

Material	Surface cutting speed (m/min)	Linear feed rate (mm/rev)
Aluminium	80 to 150	0.2 to 1.00
Brass	40 to 80	0.2 to 1.00
Cast iron	15 to 30	0.15 to 0.7
Copper	20 to 40	0.2 to 1.00
Mild steel	20 to 40	0.2 to 1.00

(HSS) cutting tool is used. Faster rates can be achieved with silicon carbide cutting tools.

Example 10.9

An alloy tube having an outside diameter of 32mm is to be turned on a lathe which has a spindle speed of 360 RPM. Determine the cutting speed.

The cutting speed can be calculated from:

$$V = \frac{\pi DS}{1000} \text{ m/min}$$

where $D = 32$mm and $S = 360$ rev/min.

Thus:

$$V = \frac{3.142 \times 32 \times 360}{1000} = 36.2 \text{m/min}$$

Example 10.10

Calculate the spindle speed in rev/min for turning a 25mm diameter workpiece at a cutting speed of 30m/min.

Now the spindle speed is given by:

$$S = \frac{1000V}{\pi D} \text{ rev/min}$$

where $V = 30$m/min and $D = 25$mm.

Thus:

$$S = \frac{1000 \times 30}{3.142 \times 25} = 382 \text{ rev/min}$$

Example 10.11

In the previous example, how long will it take to cut a 50mm length of the workpiece if the feed rate is 0.5mm/rev?

The feed rate can be calculated from:

$$F = f \times S \text{ mm/min}$$

where $f = 0.5$mm/rev and $S = 382$rev/min.

Example 10.11 (Continued)

Thus:

$$F = 0.5 \times 382 = 191 \text{ mm/min}$$

In order to determine the time needed to cut a 50mm length of the workpiece we simply need to apply the relationship:

$$t = \frac{d}{F}$$

where t = time taken, d = distance moved and F = feed rate

In this case we have d = 50mm and F = 191mm/min.

Thus:

$$t = \frac{d}{F} = \frac{50}{191} = 0.262 \text{ min.} \approx 16 \text{ sec}$$

Test your knowledge 10.20

A mild steel bar having a diameter of 60mm is to be turned on a lathe which has been set to a spindle speed of 400 rev/min. Determine the cutting speed. Is this within the range that you would expect?

Test your knowledge 10.21

Determine the feed rate expressed in mm/min if it takes 2 minutes to turn an aluminium bar of length 220mm when the spindle speed of a lathe has been set to 240 rev/min.

Learning outcome 10.11

Construct simple graphs

Graphs provide us with a useful way of representing data in visual form. They can also be used to show, in a simple pictorial way, how one *variable* affects another variable. Many different types of graph are used in engineering. We shall start by looking at the most basic type, the straight-line graph.

Straight-line graphs

Earlier in Section 10.5 we introduced the idea of *proportionality*. In later sections you saw numerous examples of this. For example, in Section 10.10 you saw how, when turning, the cutting speed, V, is directly proportional to the spindle speed, S. We could have explained this using a straight-line graph from which you could read off corresponding values of V and S, as shown in Figure 10.21.

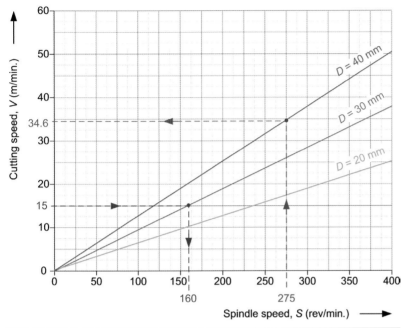

Figure 10.21 Graphs showing cutting speed, V, plotted against spindle speed, S, for a turning operation.

In Figure 10.21 we've shown three separate lines on the graph. The green line relates to a workpiece having a diameter of 20mm, while the blue and red lines relate to a 30mm and 40mm diameter workpiece, respectively. These graphs can be very useful. For example, if we need to find the spindle speed corresponding to a cutting speed of 15m/min and a 30mm diameter workpiece we can simply look across to the blue line and then look down to the spindle speed axis to read off the corresponding spindle speed (160 rev/min). Alternatively, if we need to find the cutting speed for a 40mm diameter workpiece when the spindle speed is set to 275 rev/min, we just need to look up from a spindle speed of 275 rev/min to the red line and then look across to the cutting speed axis to arrive at a cutting speed of 34.6m/min.

Test your knowledge 10.22

Use the graphs shown in Figure 10.21 to determine:

1 The spindle speed when a cutting speed of 40m/min is used when turning a workpiece having a diameter of 40mm.

2 The cutting speed corresponding to a spindle speed of 225 rev/min when turning a workpiece having a diameter of 30mm.

Plotting data graphically

Graphs can also be very useful when plotting results of a series of measurements. For example, let's assume that we have obtained the following values in an experiment to measure corresponding values of voltage, V, and current, I, flowing in an electric circuit:

Voltage, V (V)	0	1	2	3	4	5	6
Current, I (A)	0	0.33	0.66	1.0	1.33	1.66	2.0

This data can be plotted in the form of a graph, as shown in Figure 10.22. To obtain the graph, a point is plotted for each pair of corresponding values for V and I. When all the points have been drawn they are connected together by drawing a line. Notice that, in this case, the line that connects the points together takes the form of a straight line. This is *always* the case when two variables are directly proportional to one another.

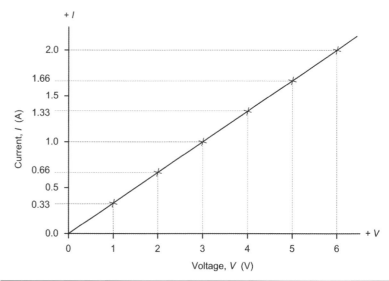

Figure 10.22 Graph of current, I, plotted against voltage, V, for an electric circuit.

It is conventional to show the *dependent variable* (in this case it is current, I) plotted on the vertical axis and the *independent variable* (in this case it is voltage, V) plotted on the horizontal axis. If you find these terms a little confusing, just remember that, what you know is usually plotted on the horizontal scale while what you don't know (and may be trying to find) is usually plotted on the vertical scale. In fact, the graph contains the same information regardless of which way round it is drawn! Now let's look at another example:

Example 10.12

The following measurements are made on an electronic component:

Temperature, θ (°C)	10	20	30	40	50	60
Resistance, R (Ω)	105	110	115	120	125	130

Plot the graph showing how the resistance of the component varies with temperature. Determine the resistance of the component at 0°C and suggest the relationship that exists between resistance and temperature.

The results of the experiment are shown plotted in graphical form in Figure 10.23. Note that the graph consists of a straight line but that it does not pass through the *origin* of the graph (i.e. the point at which θ and V are 0°C and 0V respectively). The second most important feature to note (after having noticed that the graph is a straight line) is that, when $\theta = 0°C$, $R = 100Ω$.

By looking at the graph we could suggest a relationship (i.e. an *equation*) that will allow us to find the resistance, R, of

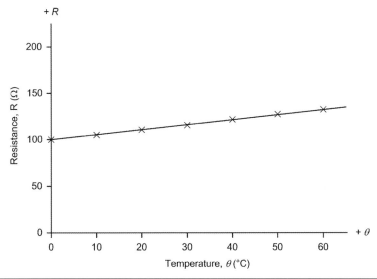

Figure 10.23 See Example 10.12.

Key point

Graphs provide us with a way of visualizing data. When a graph is plotted the independent variable is usually plotted on the horizontal axis (the x-axis) whilst the dependent variable is usually plotted on the vertical axis (the y-axis).

Example 10.12 (Continued)

the component at any given temperature, θ. In this case the relationship is simply:

$$R = 100 + \frac{\theta}{2} \ \Omega$$

If you need to check that this works, just try inserting a few pairs of values from those given in the table. You should find that the equation balances every time!

Learning outcome 10.12

Calculate values for similar triangles

As you saw earlier in Section 10.6, a triangle is an object that has three sides. The sum of the interior angles of a triangle is 180°. Thus, if we know any two of the angles of a triangle we can find the remaining (third) angle.

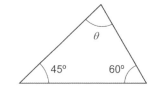

Figure 10.24 See Example 10.13.

Example 10.13

Determine the third angle of the triangle shown in Figure 10.24.

Now the sum of the angles in a triangle is 180°. Hence:

$\theta + 45° + 60° = 180°$

So $\theta = 180° - 45° - 60° = 180° - 105° = 75°$

Some shapes involving triangles are shown in Figure 10.25. In Figure 10.25a the shape is a triangle which is half of a perfect square having sides 1m × 1m. The area of the triangle is thus $0.5 \times 1 \times 1 = 0.5 m^2$.

In Figure 10.25b the area is made up from three perfect squares (each of area $1m^2$) and one triangle having an area of $0.5m^2$. The total area is thus $3 + 0.5 = 3.5m^2$.

Finally, the shape in Figure 10.25c can be divided into a rectangle with an area of $12m^2$ and a triangle having an area of $3m^2$. The total area is thus $12 + 3 = 15m$.

Key point

The sum of the angles of a triangle is 180°. Thus, if two of the angles are known the third can easily be found.

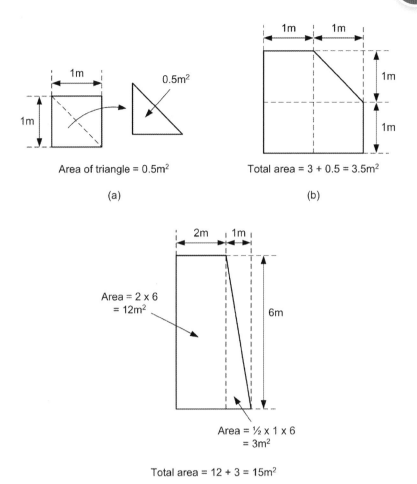

Area of triangle = 0.5m²

(a)

Total area = 3 + 0.5 = 3.5m²

(b)

Area = 2 x 6
= 12m²

Area = ½ x 1 x 6
= 3m²

Total area = 12 + 3 = 15m²

(c)

Figure 10.25 Some shapes involving triangles.

Test your knowledge 10.23

Figure 10.26 shows the arrangement used in a mobile crane. Find the three unknown angles, *A*, *B* and *C*.

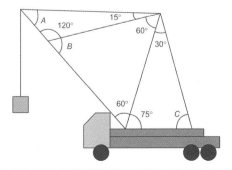

Figure 10.26 See Test your knowledge 10.23.

Test your knowledge 10.24

Figure 10.27 shows the dimensions of the tail plane of a *Gulfstream* aircraft. Determine the total surface area of one side of the tail plane (for the purposes of this exercise you can assume that the tail plane is perfectly flat). Show all working in your answer and include a diagram showing how you divided the shape into a rectangle and two triangles.

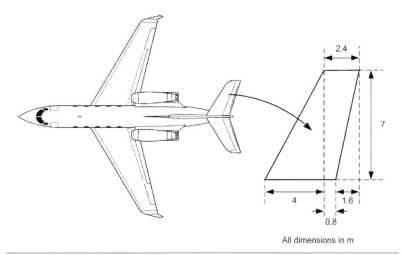

All dimensions in m

Figure 10.27 See Test your knowledge 10.24.

Test your knowledge 10.25

Find the area of each shape shown in Figure 10.28.

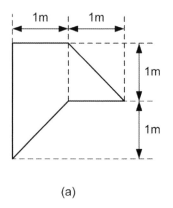

(a)

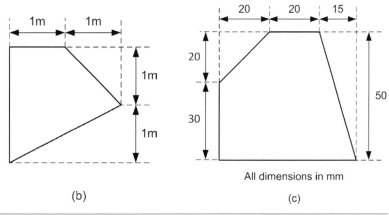

(b)

(c)

All dimensions in mm

Figure 10.28 See Test your knowledge 10.25.

Similar triangles

Triangles are said to be *similar* when they have the same three angles (in which case their three sides will be in direct proportion). Figure 10.29 shows two similar triangles. Note how, even though the lengths of their sides are different, they both share the same three angles, where angle $A = P$, $B = Q$, and $C = R$. Provided you know that two triangles are similar, you can determine the lengths of any unknown sides by simple ratio, as in the following example.

$$\frac{a}{p} = \frac{b}{q} = \frac{c}{r}$$

Similar triangles might sometimes be hard to spot. Figure 10.30a and 10.30b are not oriented the same way, but nevertheless they are similar. By rotating the triangle in Figure 10.30a you can arrive at the triangle shown in Figure 10.30c. This now looks similar to the triangle shown in Figure 10.30b. These two triangles have the same angles, 30°, 60° and 90° (recall that these should add up to 180°) and their sides are in the ratio 1.5:1.

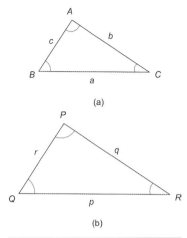

Figure 10.29 Two similar triangles.

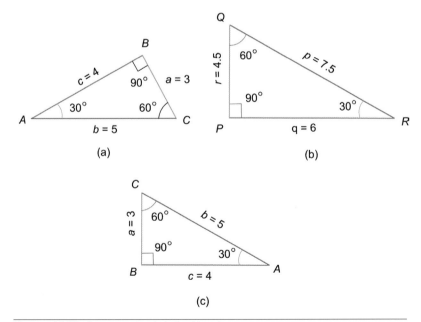

Figure 10.30 An example of similar triangles.

Test your knowledge 10.26

In Figure 10.31 two of the triangles are similar. Identify these two triangles and determine the missing angles and lengths.

Key point

Similar triangles have the same angles and the lengths of their sides are in direct proportion.

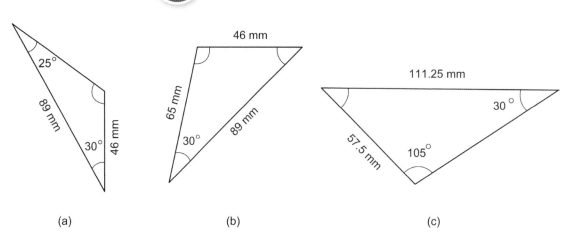

(a)   (b)   (c)

Figure 10.31 See Test your knowledge 10.26.

Learning outcome 10.13

Identify the techniques used for calculating approximation

Because precise calculations can take time (and may not always be required), engineers frequently need to make quick estimates based on approximate values. It's also quite useful to make a rough estimate of an answer to a problem or calculation before you arrive at a more accurate answer using a calculator. If there is a big discrepancy between the approximate result and the calculated answer you will know that something must have gone wrong and you will need to check your working again!

You may sometimes hear approximate values referred to as *ballpark figures*. Approximation is frequently used when we don't actually need an exact figure straight away (we can always work this out later). Instead, we just need to get a feel for what the exact value would be.

Let's suppose that we need to estimate the cost of a rectangular sheet of metal with sides 2.9m and 4.1m and we know that the metal costs £1.51 per square metre. To arrive at an approximate estimate of the cost, we could simply round the numbers up or down and then multiply them together in order to find the area (in square metres) and then multiply the result by 1.5 (instead of 1.51) in order to arrive at an approximate cost in pounds.

Hence estimated area of metal square metres = 3 × 4 = 12 square metres.

Estimated cost = 12 × £1.5 = £18

We can do all of this by applying a little mental arithmetic – there's actually no need to use a calculator!

Now, if we needed to arrive at an exact value, we could use the same reasoning but enter the values into a calculator (or use long multiplication) as follows:

Actual area of metal square metres = 2.9 × 4.1 = 11.89 square metres.

Actual cost = 11.89 × £1.51 = £17.594

As you can see, our estimate of the cost was actually quite close to the real value but we got there much more quickly! Simply rounding up or rounding down to the nearest whole number will often provide you with a very quick estimate.

> **Key point**
>
> Approximation is used whenever we need a rough estimate of a value. Approximation is also a good way of checking the validity of a complicated calculation using a calculator!

Test your knowledge 10.27

1 A race car travels at 204km/h around a test circuit which has a total length of 8.1km. Estimate (without using a calculator) how many circuits the car will complete in one hour?

2 Determine the actual number of circuits completed by the race car in Question 1.

3 A coil-winding machine rotates at 122 rev/min. Approximately, how long will it take to manufacture a coil having 596 turns?

Review questions

1. Express a) 5¾ and b) 2⅔ as decimal values.

2. Express a) 0.8 and b) 0.625 as fractional values.

3. Find x if $x = \dfrac{2.25 \times 5.5}{4.75}$

4. Express a) 15cm in m and b) 27.5kN in N.

5. Determine the mean, median and mode of the following ten numbers: 20, 16, 19, 20, 20, 19, 21, 17, 18 and 22.

6. A gearbox has a ratio of 8:1. If the input shaft turns at 120 rev/min. How many turns will the output shaft make in a time interval of 4 minutes?

7. Calculate the total surface area of a cube having sides of 0.5m.

8. A right-angled triangle has two perpendicular sides measuring 150mm and 200mm. Determine the area of the triangle.

9. Calculate the diameter of a circular disc if it has an area of 12cm².

10. A cylindrical storage tank has a diameter of 2.5m and a height of 4m. Determine the capacity of the tank in cubic metres.

11. Determine the weight of a machined part having a mass of 1.25kg.

12. In a box of 250 batteries, 50 are known to be faulty. If a battery is randomly selected from this box, what is the probability that it will be working?

13. Find x if $x = \dfrac{\sqrt{5.5^2 - 14.25}}{32}$

14. Make a the subject of the equation, $v = u + at$

15. Make R the subject of the equation, $Z = \sqrt{R^2 + X^2}$

16. A lathe operates with a spindle speed of 180 rev/min. Determine the cutting speed if the lathe is to be used for turning a workpiece having a diameter of 32mm.

17. The following data is obtained in an experiment on a belt conveyor:

Time, t (s)	0	2	4	6	8	10
Distance, d (m)	2.5	2.9	3.3	3.7	4.1	4.5

Plot the data as a graph. Label your axes clearly and use the graph to determine a) the distance at a time of 5 seconds, and b) the time taken for the conveyor to cover a distance of 4.25m.

18. Two angles of a triangle are 35° and 55°. What will the third angle be and what special property will the two shorter sides of the triangle have?

19. A vehicle travels at a constant speed of 2.48m/s. Approximately, how long will it take for the vehicle to travel 98m?

20. A fuel tank is 70% full. If the tank has a total capacity of 1100 litres, estimate the amount of additional fuel required to fill the tank.

Chapter checklist

Learning outcome	Page number
10.1 Add, subtract, multiply and divide: whole numbers, fractions and decimals.	246
10.2 Convert fractions to decimals and decimals to fractions.	249
10.3 Identify the metric and imperial systems and the preferred standard form.	250
10.4 Calculate average, mean, median and mode.	254
10.5 Calculate ratio, proportion and percentages.	255
10.6 Calculate area, surface area, mass, volume, capacity.	258
10.7 Calculate probability.	267
10.8 Calculate the square and square root of a number.	269
10.9 Transpose simple formulae.	270
10.10 Calculate speeds and feeds.	272
10.11 Construct simple graphs.	276
10.12 Calculate values for similar triangles.	280
10.13 Identify the techniques used for calculating approximation.	284

Engineering science

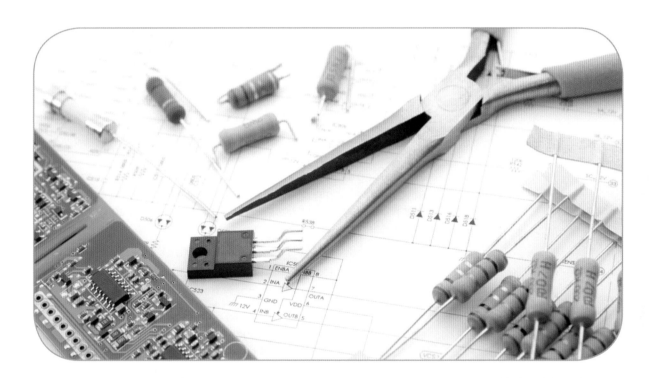

Learning outcomes

When you have completed this chapter you should understand fundamental science applied to engineering, including being able to:

11.1 Recognize common SI units.

11.2 State the types of forces used in engineering.

11.3 Calculate moments and levers.

11.4 Calculate heat input and change in length.

11.5 Identify the modes of heat transfer.

11.6 List the causes of friction.

11.7 Identify how and why materials are selected with low frictional values.

11.8   Identify structures and states of matter.

11.9   Recognize the main principles of the basic theory of electricity.

11.10 Perform simple calculations using the basics of electricity.

11.11 Calculate resistors in series and parallel circuits.

11.12 Identify lines of flux within magnetic fields.

11.13 Recognize the relationship between conductors, current, magnetic fields and relative movement.

Chapter summary

Engineers are problem solvers. As well as knowing *how* things work, they need to know *why* they work and *how* to make them work better. In order to do this they need to understand the science that underpins all branches of engineering and be able to apply that knowledge in their everyday work.

Engineering science bridges the gap between theoretical science and practical engineering. A key feature of engineering science is that it uses scientific principles and mathematics to solve real-world engineering problems, such as:

- how to reduce wear on the moving parts of a mechanism
- how to minimize the effort required to raise a load
- how to efficiently convert energy from one form to another.

As with the previous chapters, this chapter takes a topic-based approach and each of its sections is devoted to a specific learning outcome. Individual sections deal with units, forces, moments and levers, heat, friction and the different states of matter. The chapter will also introduce you to electricity and how to solve simple series and parallel circuits. The chapter concludes by looking at magnetism and electromagnetism, including some examples that show how engineering science provides solutions to some common problems. This chapter will also provide you with an opportunity to further develop your skills in mathematics.

Learning outcomes 11.1

Recognize common SI units

You will find that quite a large number of units and symbols are used in science and engineering so let's get started by introducing some of them. In fact, it's important to get to know these units and also to be able to recognize their abbreviations and symbols before you actually need to use them on a regular basis. Later we will go on to perform calculations and solve problems using these units, but for now we will simply list some of the most common units and symbols so that at least you can begin to know something about them.

Units and symbols

The units shown in Table 11.1 are called *fundamental units* (or *base units*) and they are part of the International System (known as 'SI') of units. Other units can be derived from the seven fundamental units. These are called *derived units* and a selection of them is shown in Table 11.2.

Key point

All derived units are derived in terms of the seven fundamental (or base) units.

Table 11.1 Fundamental SI units.

Name	Symbol	Unit	Unit abbreviation
Mass	M	kilogram	kg
Length	L	metre	m
Time	T	second	s
Electric current	I	ampere	A
Temperature	θ	kelvin	K
Amount of substance	N	mole	mol
Luminous intensity	J	candela	cd

Key point

Symbols used for electrical and other quantities are normally shown in italic font while units are shown in normal (non-italic) font. Thus *m* and *v* are symbols whilst J and W are units.

Table 11.2 Some common derived units.

Quantity	Unit	Abbreviation	Derivation
Energy (or *work*)	joule	J	1J of energy (or *work* done) is used when 1N of force moves through 1m in the direction of the force
Force	newton	N	Unit of force (a force of 1N gives a mass of 1kg an acceleration of 1m/s²)
Pressure	pascal	Pa	A pressure of 1Pa exists when a force of 1N is exerted over an area of 1m²
Power	watt	W	A power of 1W is developed when 1J of energy is used in 1s
Electric charge	coulomb	C	An electric charge of 1C is transferred when a current of 1A flows for 1s

(Continued)

Table 11.2 *(Continued)*

Quantity	Unit	Abbreviation	Derivation
Frequency	hertz	Hz	A wave has a frequency of 1Hz if one complete cycle occurs in 1s
Velocity	metre per second	m/s	A body travelling at a velocity of 1m/s moves through 1m every second in the direction of travel
Acceleration	metre per second per second	m/s²	A body travelling with an acceleration of 1m/s² increases its velocity by 1m/s every second in the direction of travel

Test your knowledge 11.1

1 What are the SI units for: a) mass, b) temperature and c) force?
2 What symbol is used to represent electric current and what are its SI units?

Test your knowledge 11.2

1 An electric motor uses 20J of energy in half a second. At what rate is energy consumed by the motor and what power does this correspond to?
2 Acceleration is the rate at which the speed of a body changes. A military aircraft increases its speed from 250m/s to 300m/s in 5s. What acceleration is this?
3 Electric current is the rate at which charge is moved. How much charge is moved when a current of 5A flows for half a minute?

Learning outcomes 11.2

State the types of forces used in engineering

Put simply, a force is a push or pull exerted by one object on another. If the object remains in *equilibrium* (i.e. if it doesn't move or change in some way) then, for each force acting on the object there will be another equal and opposite force that acts against it. The force that is applied to an object is often called an *action* while the opposing force is referred to as a *reaction*. As long as the object doesn't move or change, action and reaction will be equal and opposite.

It also follows that, if the forces of action and reaction acting on an object are not equal and opposite, the object will move or change in some way. You can test this theory out very easily by finding a wall and pushing against it. If the wall doesn't move (hopefully it won't!) then you will experience a force pushing back. If you increase the force that you apply to the wall (the *action*) the force pushing back (the *reaction*) will also increase by the same amount.

Now try the same experiment by pushing against a door that is partially open. There will still be some force exerted back by the door but this will be much less than the force that you apply. Because of this imbalance of forces (action being greater than reaction) the door will move and will swing open. These simple experiments lead us to the following conclusions:

1 When a body is at rest (or in *equilibrium*), the *action* and *reaction* forces acting on it will be equal and opposite.
2 If the action and reaction forces acting on a body are not equal and opposite, a change (e.g. *motion*) will be produced.

Force is measured in newtons (N) where 1N is defined as the force required to accelerate a mass of 1kg at a rate of 1m/s^2.

Weight

Weight is an example of a force. We already know from Section 10.6 that since the weight of a body is determined by its mass and the gravitational force acting on it, mass and weight are closely related. Weight can be calculated from the relationship:

$$W = mg \text{ Newton}$$

Where W is the weight in newtons (N), m is the weight in kg, and g is the gravitational acceleration (in m/s^2). On the surface of the Earth, g is a constant equal to 9.81m/s^2. Thus, on Earth:

$$W = 9.81 \times m$$

On the surface of the moon, g takes the much lower value of a constant equal to 1.622m/s^2. This is about one-sixth of that on the Earth. Thus a person weighing 800N on Earth would weigh a mere 133N on the moon.

Example 11.1

A light alloy beam has a mass of 17.5kg. Determine the weight of the beam.
Here we will assume that the beam is being used on the surface of the Earth, in which case:

$$W = mg = 17.5 \times 9.81 = 171.68N$$

Density

The density of a body is defined as the mass per unit volume. In other words, the density of an object is found by dividing its mass by its volume. Some materials are more dense than others. For example, a block of steel is more dense than a block of wood. The density of a particular material is a fundamental property of that material. Expressing this as a formula:

$$\text{Density, } \rho = \frac{m}{V} \text{ kg} / \text{m}^3$$

where m is the mass in kg and V is the volume in m^3.

We sometimes express the density of an object relative to that of pure water (at 4°C). The density of water under these conditions is 1000kg/m^3. The densities (and *relative densities*) of various engineering materials are shown in Table 11.3.

Table 11.3 Density of various materials.

Material	Density (kg/m3)	Relative density
Aluminium	2700	2.7
Brass	8500	8.5
Cast iron	7350	7.35
Concrete	2400	2.4
Copper	8960	8.96
Glass	2600	2.6
Mild steel	7850	7.85
Wood (oak)	690	0.69

Example 11.2

Determine the mass of an aluminium block which has the following dimensions:

$$0.05\text{m} \times 0.11\text{m} \times 0.275\text{m}$$

The volume of the aluminium block (see Section 10.6) can be calculated from:

$$V = 0.05 \times 0.11 \times 0.275 = 0.00151\text{m}^3$$

From Table 11.4 the value of ρ for aluminium is 2700kg/m^3. The mass of the block will thus be:

$$m = 2700 \times 0.00151 = 4.08\text{kg}$$

Test your knowledge 11.3

1 A sample of a metal alloy has a volume of $0.02m^3$ and a mass of 60kg. What is the density of the material and what would it weigh on the surface of the moon?
2 Determine the weight of a solid copper bar having a length of 0.2m and a cross-sectional area of $4cm^2$.

Properties of a force

Every force has three important properties that are used to describe it. These properties are:

- size (or *magnitude*), see Figure 11.1a
- direction of action, see Figure 11.1b
- point of application, see Figure 11.1c.

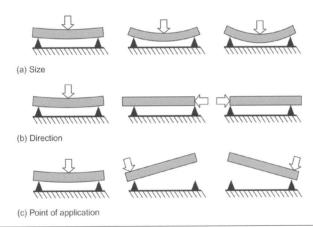

(a) Size

(b) Direction

(c) Point of application

Figure 11.1 Properties of a force.

Pressure

Pressure (or stress) is exerted whenever a force is applied to an object such as a floor, wall or the surface of a container. Pressure is defined as the ratio of the force (or *load*) applied perpendicular (i.e. at right angles) to the surface area over which the load acts. Pressure is measured in pascals, Pa. Thus:

$$\text{Pressure, } P = \frac{F}{A} \text{ pascals}$$

where F is the force in N and A is the area over which it is applied in m^2.

Key point

Weight is an example of force. The weight of a body is the product of its mass and acceleration caused by gravity.

Key point

The density of a body can be found by dividing its mass by its volume.

Key point

Pressure is exerted when a force is applied to an object such as a floor or wall. Pressure can be calculated from the ratio of perpendicular force to the area over which the force acts.

Example 11.3

A machine weighs 12.5kN. Determine the pressure exerted on a workshop floor if the weight of the machine is distributed over a surface area of 0.875m^2.

The pressure on the floor can be calculated from:

$$P = \frac{12,500}{0.875} = 14,290 \, Pa = 14.29 \, kPa$$

Test your knowledge 11.4

A mild steel girder has a mass of 560kg and is supported by two concrete pillars, each having a cross-sectional area of 0.044m^2. Assuming that the load is distributed evenly, determine the pressure acting on each of the pillars.

Learning outcomes 11.3

Calculate moments and levers

Moments

A moment is simply a force that produces a *turning effect*. The magnitude of a moment depends on the product of the force applied and the perpendicular distance from the pivot or axis to the line of action of the force. Hence:

$$M = F \, d \text{ newton-metres}$$

where M is the moment in newton-metres, F is the force in newtons, and d is the distance in metres. Note that if a load is expressed as a mass (in kg) it will be necessary to determine the corresponding force in newtons by multiplying the mass by 9.81.

Being able to exert more force on an object by increasing the distance at which it is applied from the pivot point leads us to a useful device known as a lever. You might have already used a lever when changing a tyre or when removing the lid from a container. Turning moment is proportional to distance, thus, to gain more leverage you simply need to increase the length of the lever.

Example 11.4

The diagram shown in Figure 11.2 shows a spanner exerting a force on a nut (note that, in order to minimize the force applied to the spanner we have applied it at right angles to the spanner's major axis). Determine the turning effect on the nut.

The turning effect (at right angles to the spanner) will be the product of the applied force, F, and the length of the spanner, d.

Thus turning moment $= F \times d = 50 \times 0.2 = 10\,\text{Nm}$.

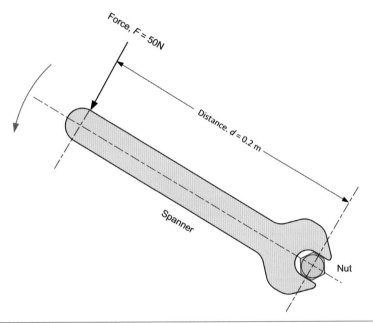

Figure 11.2 See Example 11.4.

When several forces are present

In Example 11.4 the result of applying force to the end of the spanner is that the nut will move and both nut and spanner will rotate. In this example only one force is present but in many other cases more than one force might be acting but, instead of producing motion, we might actually want to keep things stationary. Thus, when several forces are acting at the same time we often need to determine the force that, when applied, will maintain *equilibrium*. Recall that, when a system of forces is in *equilibrium* there will be no movement and the total clockwise moment (CW) will be equal to the total anticlockwise moment (ACW).

Example 11.5

Figure 11.3 shows a beam of negligible mass having an overall length of 7m. The beam is supported by a pivot point which is 3m from the left-hand end and 4m from the right-hand end. On the right of the pivot, forces of 10N and 5N are applied at distance of 1m and 4m respectively. On the left of the pivot an unknown force, F, is applied 3m from the pivot. The direction of action of each of these forces has been shown on the diagram. Determine the value of the unknown force, F, in order to preserve equilibrium.

Since the system is in equilibrium we can infer that the total clockwise (CW) moment is equal to the total anti-clockwise (ACW) moment.

The total CW moment is: $(10 \times 1) + (5 \times 4) = 30\text{Nm}$.

The total ACW moment is simply: $F \times 3 = 3F$ Nm.

Equating the CW and ACW moments gives:

$$3F = 30\text{Nm}$$

From which $F = 10\text{N}$.

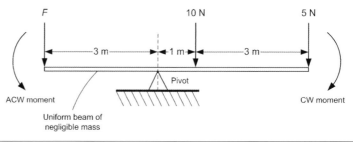

Figure 11.3 See Example 11.5.

Figure 11.4 A crane jib is an example of a beam. Note the adjustable counterweight made from concrete blocks.

Test your knowledge 11.5

Find the unknown force, F, required to preserve equilibrium in Figure 11.5.

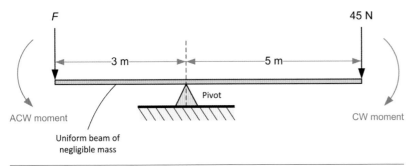

Figure 11.5 See Test your knowledge 11.5.

Learning outcomes 11.4

Calculate heat input and change in length

Heat is something that we are very familiar with; we quickly know whether something is hot or cold just by touching it. Heat is a form of energy that is transferred within the individual particles that make up a particular substance. Some substances, like iron and steel, convey heat very quickly. Others, like wood and plastic, are much slower to convey heat.

Heat and temperature are different things. Temperature does not depend on the size of a body nor does it depend on what it is made of. The amount of heat energy, on the other hand, will depend on several factors, including the type of material and how much of it is present. For example, a cup of tea might be at exactly the same temperature as the water in a bath, but the bath water will have considerably more heat energy due to its much larger volume.

Heat energy can be transferred from one object to another. This transfer of energy is due to the difference in temperature between the two objects. The amount of heat transferred is usually denoted by the symbol Q, and it is measured in Joules, J.

Key point

Heat is a form of energy that is transferred between the individual particles that make up a particular substance. Temperature is a measure of the average heat of these particles. This is not the same as the total amount of *energy* present.

Temperature

You will already be familiar with temperature measured on the Celsius scale where 0°C and 100°C respectively correspond to the freezing and boiling points of water. The absolute zero temperature

11 Engineering mathematics and science principles

point occurs at a temperature of –273.15°C. At this point all molecular vibrations within a material will cease. In the SI system (see Table 11.1) we mentioned that temperature is measured on the Kelvin (K) scale. This scale starts from absolute zero, so 0K = –273°C and, since the intervals on the Kelvin scale are the same as those on the Celsius scale, 273K = 0°C and 373K = 100°C. Thus, to convert from Celsius to the Kelvin scale we simply add 273. In the USA (and in several other countries) temperature is measured on the Fahrenheit scale where –32°F corresponds to 0°C and 212°F corresponds to 100°C, as illustrated in Figure 11.6. To convert between degrees Fahrenheit and degrees Celsius you can use the relationships given in Table 11.4.

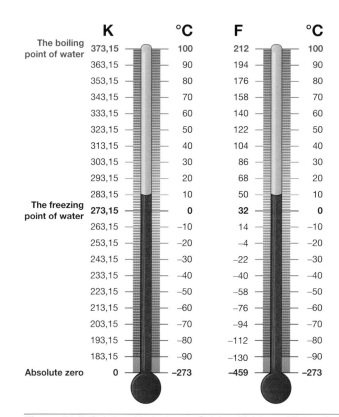

Figure 11.6 Relationship between Celsius, Kelvin and Fahrenheit temperature scales.

Table 11.4 Converting between degrees Fahrenheit and degrees Celsius.

To convert from:	Degrees Celsius to degrees Fahrenheit	Degrees Fahrenheit to degrees Celsius
Relationship to use:	$F = \left(C \times \dfrac{9}{5} \right) + 32$	$C = \left(F - 32 \right) \times \dfrac{5}{9}$

Example 11.6

Convert a temperature of 70 degrees Fahrenheit to Celsius and Kelvin.

To convert from Fahrenheit to Celsius:

$$°C = \left(68 - 32\right) \times \frac{5}{9} = 36 \times \frac{5}{9} = 20$$

To convert this Celsius value to Kelvin we simply need to add 273 to the result:

Thus 70°F = 20°C = 293K.

Key point

In the SI system temperature is measured in Kelvin where 273K = 0°C and 373K = 100°C.

Test your knowledge 11.6

1 Convert 150°C to °F.
2 Convert 20°F to °C.
3 Convert 305°K to °F.

Thermal expansion

Solids, liquids and gases all expand with an increase in temperature. This *thermal expansion* is dependent on the nature of the material and the magnitude of the temperature increase. We normally measure the linear expansion of solids, i.e. the increase in length of a bar of the material.

Every solid material has a linear *expansivity*. This is the amount the material will expand in metres per Kelvin or per degree Celsius. This value is often referred to as the *coefficient of linear expansion* (α). Some typical values for α are given in Table 11.5.

Table 11.5 Coefficient of linear expansion, α, for various materials.

Material	Coefficient of linear expansion, α (/°C)
Aluminium	0.000024
Brass	0.000019
Copper	0.000017
Invar	0.0000015
Iron (cast)	0.000010
Steel	0.000012

Change in length

Given the length of a material, l, its coefficient of linear expansion, α, and the temperature rise, Δt, the increase in its length can be calculated using:

Increase in length, $\Delta l = \alpha l\,(t_2 - t_1)$

Notice that we've used the Greek symbol 'Δ' to mean 'a change'.

Example 11.7

A steel bar has a length of 4m at 10°C. By how much will the length of the bar increase when heated to 350°C and what will its new length be?

The increase in length can be calculated from:

$$\Delta l = \alpha l\,(t_2 - t_1)$$

where $\alpha = 0.000012$ (see Table 11.5), $l = 4$m, $t_2 = 350$°C and $t_1 = 10$°C.

Thus:

$$\Delta l \times 4 \times (350 - 10) = 0.01632\text{m} = 16.32\text{mm}$$

The new length of the bar will be:

$$l = l + \Delta l = 4 + 0.01632 = 4.01632\text{m}$$

Test your knowledge 11.7

An alloy beam has a length of 2.4m at a temperature of 20°C. If the alloy has a coefficient of linear expansion of 0.000024/°C, determine the change in length when the alloy is heated to a temperature of 420°C.

Learning outcomes 11.5

Identify the modes of heat transfer

Heat can be transferred in three different ways: conduction, convection and radiation. Technically, only conduction and radiation are true heat transfer processes, because they both depend totally and utterly on a temperature difference being present. Nevertheless, since convection can accomplish transmission of energy by moving a mass of liquid or gas from a high to a low temperature region, it can also be regarded as a heat transfer mechanism. We will briefly describe each one of these three processes:

Conduction

Conduction occurs when two bodies having different temperatures are in contact with one another. In this situation heat energy will be transferred from the warmer body to the colder body. This transfer of energy will continue until the two bodies reach the same temperature (the warmer body will cool down while the colder body will warm up). Conduction works well when two solid materials are in direct contact with one another.

Convection

Convection occurs in liquids and gases where there is movement of the liquid or gas from a warmer region to a cooler one. When this happens, cooler liquid or gas moves to take the place of the warm liquid or gas that has moved. The cycle of warming and cooling becomes continuous and results in continuous motion within the liquid or gas as energy is transferred from the warmer to the colder regions. Since the gap between adjacent particles in a liquid or gas is appreciably larger than that within solid materials, convection is the most efficient means of heat transfer within liquids and gases.

Radiation

Unlike conduction and convection, radiation doesn't need a medium through which heat can be transferred. Thus, radiation can be transferred through empty space, as witnessed by the fact that we can warm our bodies by exposing them to the sun. Radiation is due to energy levels within a cold body becoming raised due to the presence of electromagnetic waves generated by a hot body. When striking the cold body, the emitted radiation is either absorbed by, reflected by or transmitted through the body.

Key point

Heat energy can only be transferred from a hot body to a cold body. If two bodies are at the same temperature no heat will be transferred between them.

Key point

Heat energy can be transferred by conduction, convection or radiation.

Specific heat capacity

From what we've just said about heat transfer, it will be apparent that different materials have different capacities for absorbing and transferring heat. The thermal energy needed to produce a temperature rise in a particular material depends on the mass of the material, the type of material and the temperature rise to which the material is subjected.

The inherent ability of a material to absorb heat for a given mass and temperature rise is dependent on the material itself. This property of the material is known as its specific heat capacity. The specific heat capacity of a material is the thermal energy required to produce a one degree Kelvin increase in temperature of a mass of 1kg. Therefore knowing the mass of a substance and its specific heat capacity, it is possible to calculate the amount of heat energy required to produce any given temperature rise, from:

Thermal energy, $Q = m\,c\,\Delta t$ joules

where c = specific heat capacity of the material (J per kg.K) and Δt is the temperature change.

Some typical values of specific heat capacity are given in Table 11.6. Once again, notice how we've used the Greek symbol 'Δ' to mean 'a change'.

Table 11.6 Specific heat capacity, c, for various materials.

Material	Specific heat capacity, J/kg.K
Aluminium	897
Copper	380
Glass	840
Iron	450
Steel	490
Water	4181

Example 11.8

How much thermal energy will be required to raise the temperature of 5kg of aluminium from 23°C to 50°C?

The required thermal energy can be calculated from:

$$Q = m\,c\,\Delta t \text{ J}$$

where m = 5kg, c = 897 (see Table 11.6), and Δt = (50 – 23) = 27°C.

Thus:

$$Q = 5 \times 897 \times 27 = 12{,}1095\text{J} = 121\text{kJ}$$

Test your knowledge 11.8

A process produces 20kJ of thermal energy. What temperature change will be produced in 2kg of aluminium if all of the heat energy is transferred to it?

Learning outcomes 11.6

List the causes of friction

When one surface is moved over another with which it is in contact, a resistance is set up to the motion. The amount of resistance will depend on the materials concerned as well as the force that holds the two surfaces together. This resistance to movement is known as *friction.* Note that a slightly greater force (*static friction*) is required to start two surfaces moving over one another compared with the force required to keep them moving (*sliding* or *kinetic friction*).

Figure 11.7 shows a block moving over a horizontal surface. Here, friction will occur between the lower surface of the block and the upper surface of the horizontal plane. If the system shown in Figure 11.7 is in equilibrium (i.e. just on the point of moving) we can equate the forces horizontally and vertically in order to obtain the following equations:

Horizontal: $P = F = \mu N$ (where μ is the coefficient of friction)

Vertical: $N = W = mg$ (where g = 9.81m/s²)

Combining these two equations gives:

$$F = \mu N = \mu m g$$

and so:

$$\mu = \frac{F}{N}$$

Thus μ is the ratio of the force required to produce motion, F, and the force that is effectively pressing the two materials together, N.

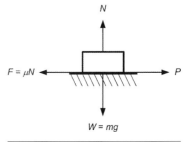

Figure 11.7 Friction and reaction.

Example 11.9

Calculate the horizontal force required to move a crate having a mass of 100kg over a floor if the coefficient of friction, μ, between the crate and the floor is 0.35.

The force required to move the crate will be given by:

$$P = F = \mu m g = 0.35 \times 150 \times 9.81 = 515N$$

Test your knowledge 11.9

What force would be required to move a load of 1250kg if the coefficient of friction between the load and the surface on which it is placed is 0.68?

Key point

Friction is the resistance to motion when one surface moves over another surface. Friction always opposes the motion that causes it.

Key point

Static friction (the force required to start an object moving) is slightly greater than sliding friction (the force that opposes motion once it has started).

Learning outcomes 11.7

Identify how and why materials are selected with low frictional values

Coefficient of friction

We've already explained that the coefficient of friction, μ, is the ratio of the force of friction between two bodies and the force that's pressing them together. It's important to note that μ is not a property of a particular material but rather it is to do with the way that two surfaces interact.

Some combinations of different materials exhibit relatively low coefficients of friction while others have high values of μ. For example, Teflon on steel has a very low coefficient of friction while rubber on concrete exhibits a very high coefficient of friction. Values of μ range from just above zero to greater than one, but most metal and alloy combinations have values between 0.4 and 0.8.

Static and kinetic friction

In order to distinguish between static and kinetic friction (see Section 11.6) we add subscripts 's' and 'k' to the coefficient, as follows:

Static coefficient of friction $= \mu_s$

Kinetic (i.e. sliding) coefficient of friction $= \mu_k$

Typical coefficients of friction between the surfaces of various materials are shown in Table 11.7.

Table 11.7 Typical coefficients of friction, μ, for various materials.

Material	Static friction, μ_s	Kinetic friction, μ_k
Aluminium on steel	0.61	0.47
Brass on steel	0.51	0.44
Copper on steel	0.53	0.36
Metal on metal (lubricated)	0.15	less than 0.1
Rubber on concrete	1.0	0.8
Steel on steel	0.74	0.57
Teflon on Teflon	0.04	0.04

Effects of friction

The relatively high value of coefficient of friction that exists between rubber and concrete and tarmac provides a car tyre with a means of gripping the road. Conversely, a low coefficient of friction between two surfaces coated with Teflon ensures that the two surfaces slide easily over one another. When surfaces in contact move relative to each other, the friction between the two surfaces can produce an appreciable amount of heat. Friction between components that are intended to move (such as *gear trains*) can result in wear which may lead to performance degradation and/or damage. Friction can arise from a number of causes including roughness of surfaces as well as surface contamination and deformation. Lubrication based on oil or grease can be applied to surfaces in order to combat the effects of friction.

Test your knowledge 11.10

1 Explain the difference between static and kinetic friction.
2 Explain why Teflon is often used as a coating for the moving parts in a mechanism.

Learning outcomes 11.8

Identify structures and states of matter

All matter exists in one of three states: solid, liquid or gas. The state in which matter exists is determined by the binding force that exists between the particles (atoms or molecules) of which the material is composed.

Solids

When atoms are very closely packed together there is a force of repulsion between them. However, at a greater separation a force of attraction exists. The force that exists between atoms is balanced at a distance which is equal to one atomic diameter. In this condition the atoms remain together in a solid form (see below). A solid has a well-defined shape and volume.

Liquids

As temperature increases, the internal vibration energy of the atoms increases to a point at which the force between adjacent atoms exceeds the inter-atomic bonding force that holds the material in its

Key point

Coefficient of friction is the ratio of force that acts between two bodies preventing motion and the force that's pressing them together.

Key point

Some combinations of material surface exhibit relatively high values of coefficient of friction while others exhibit low values of coefficient of friction. Where two surfaces need to grip one another a high value of coefficient is required. Conversely, where two surfaces should slide easily over one another a low value of coefficient is required.

solid form. At the point the material changes its state into the liquid form, atoms and molecules slide easily over one another due to the increased temperature. Unless held in a container, a liquid has a volume but not an easily defined shape.

Gases

In a gas, the atoms and molecules move randomly and move apart so as to take up all the space in the containing vessel. Molecular interaction occurs only rarely (when molecules collide). A gas has no particular shape or volume but expands until it fills any vessel in which it is placed. Note that a *fluid* may be either a liquid or a gas.

Atomic structure

The nucleus of an atom consists of *protons* and *neutrons*. Protons have a small positive charge but neutrons are electrically neutral. Surrounding the nucleus, in a discrete number of energy bands, negatively charged electrons orbit the nucleus, as shown in Figure 11.8.

Atoms are electrically neutral. This means that they exhibit no net charge, neither positive nor negative. Because of this the number of electrons must be exactly matched by the number of protons. Note that electrons in the energy bands (or shells) close to the nucleus are held tightly by means of electrostatic attraction (recall that like charges repel and unlike charges attract). Electrons in the outermost shells are held less tightly.

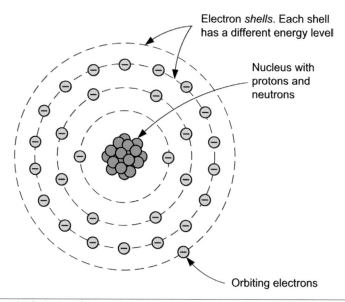

Electron *shells*. Each shell has a different energy level

Nucleus with protons and neutrons

Orbiting electrons

Figure 11.8 A simplified model of an atom.

Ions

An ion is formed when an atom gains or loses an electron. For example, if an electron is taken away from the outer shell, a positive ion will be produced.

Valency

The valence of an atom is related to the ability of an atom to enter into chemical combination with other elements. Valence is usually determined by the number of electrons in the outermost shells. These valence shells are often referred to as the *s* or *p* shells (remember that it is the outermost, most accessible shells that are the most important).

Bonding

Atoms and molecules can combine (or bond) in various ways. In *ionic* (or *electrovalent*) *bonding* one atom may donate its valence electron to a different atom, filling the outer shell of the second atom. *Covalent bonds* occur where materials share electrons between two or more atoms and *metallic bonds* where elements with a low valency (i.e. metals) give up their outer shell electrons to produce a 'sea' of mobile electrons which surround the nucleus of the atoms. This can be quite useful, as we will see in the next section.

Test your knowledge 11.11

1 What are the three states in which matter can exist?
2 What is the name given to an atom that carries a net positive charge?
3 Aluminium has three electrons in its outermost s and p shells. What is its valency?
4 An electrovalent bond is another name for what?

Key point

Matter exists in one of three forms: solid, liquid or gas.

Learning outcomes 11.9

Recognize the main principles of the basic theory of electricity

Electrical and electronic components and systems are used in a huge range of engineering applications. An aircraft, for example, would simply not get off the ground without the electrical and electronic systems that maintain the instruments, operate the flight controls, manage the engines, and provide navigation and communication with the ground. The term 'fly by wire' provides us with a clue as to just how important the role of electricity is in the operation of a modern aircraft!

Electricity is very widely used in engineering manufacture and production. Machine tools, pumps, lifts, conveyors and servos are all driven by electric motors. Electricity is used for signalling and communication and also in the automated systems that are controlled by computers.

Conductors and insulators

Electric current is the name given to the flow of electrons (or negative charge carriers). Electrons orbit the nucleus of atoms just as the Earth orbits the sun (see Figure 11.8). Electrons are held in one or more shells, constrained to their orbital paths by virtue of a force of attraction towards the nucleus, which contains an equal number of protons (positive charge carriers).

Since like charges repel and unlike charges attract, negatively charged electrons are attracted to the positively charged nucleus. A similar principle can be demonstrated by observing the attraction between two permanent magnets; the two north poles of the magnets will repel each other, while a north and south pole will attract. In the same way, the unlike charges of the negative electron and the positive proton experience a force of mutual attraction.

The outer shell electrons of a conductor can be reasonably easily interchanged between adjacent atoms within the lattice of atoms of which the substance is composed. This makes it possible for the material to conduct electricity. Typical examples of electrical conductors are metals such as copper, silver, iron and aluminium. By contrast, the outer shell electrons of an insulator are firmly bound to their parent atoms and virtually no interchange of electrons is possible. Typical examples of insulators are plastics, rubber and ceramic materials.

Electric charge and current

All electrons and protons carry a tiny electric charge but its value is so small that a more convenient unit of charge is used called the coulomb. When electric charge is made to move, the flow that results is referred to as a *current*.

You can't actually see an electric current but you might be aware of it through some of the other effects associated with it, such as heat (when a wire gets hot) or light (when a filament lamp glows) or magnetism (when iron or steel is attracted to an electromagnet).

Current is defined as the rate of flow of charge and its unit is the ampere, A. Hence:

$$I = \frac{Q}{t} \text{ amperes}$$

where Q is in coulombs, C, and t is in seconds.

So, for example, if a charge of 120C is moved in 60s the current flowing will be:

$$I = \frac{120}{60} = 2 \text{ A}$$

Key point

Charges with the same polarity (i.e. both positive or both negative) will repel one another while charges with opposite polarity (i.e. one positive and the other positive) will attract one another.

Example 11.10

How much charge will be transferred when a current of 3A flows for two minutes?

Rearranging the expression that we met earlier gives:

$$Q = I \times t \text{ C}$$

where I is in A and t is in seconds.

Thus:

$$Q = 3 \times 120 = 360C$$

Key point

Current is the rate of flow of charge. Thus, if more charge moves in a given time, more current will be flowing. If no charge moves then no current is flowing.

Test your knowledge 11.12

1 A power supply delivers a current of 15A for 10 minutes. How much charge does it deliver in this time?
2 What current is flowing when a charge of 0.45C is transferred in 2.5s?

Key point

A current of one ampere (1A) is equal to one coulomb (1C) per second.

Learning outcomes 11.10

Perform simple calculations using the basics of electricity

The ability of an energy source (e.g. a battery) to produce a current within a conductor may be expressed in terms of electromotive force (e.m.f.). Whenever an e.m.f. is applied to a circuit a potential difference (p.d.) exists. Both e.m.f. and p.d. are measured in volts (V). In many practical circuits there is only one e.m.f. present (the battery or supply) whereas a p.d. will be developed across each component present in the circuit.

The conventional flow of current in a circuit is from the point of more positive potential to the point of greatest negative potential (note that electrons move in the opposite direction!). Direct current results from the application of a direct e.m.f. (derived from batteries or a d.c. power supply). An essential characteristic of these supplies is that the applied e.m.f. does not change its polarity (even though its value might be subject to some fluctuation).

For any conductor, the current flowing is directly proportional to the e.m.f. applied. The current flowing will also be dependent on the physical dimensions (length and cross-sectional area) and material of which the conductor is composed.

The amount of current that will flow in a conductor when a given e.m.f. is applied is inversely proportional to its resistance. Resistance, therefore, may be thought of as an opposition to current flow; the higher the resistance, the lower the current that will flow (assuming that the applied e.m.f. remains constant).

Ohm's Law

Provided that temperature does not vary, the ratio of p.d. across the ends of a conductor to the current flowing in the conductor is a constant. This relationship is known as Ohm's Law and it leads to the relationship:

$$\frac{V}{I} = \text{a constant} = R$$

where V is the potential difference (or *voltage drop*) in volts (V), I is the current in amperes (A), and R is the resistance in ohms (Ω) (see Figure 11.9).

The formula may be arranged to make V, I or R the subject, as follows:

$$V = IR, \quad I = \frac{V}{R} \quad \text{and} \quad R = \frac{V}{I}$$

Figure 11.10 shows a triangle that will help you remember these three important relationships.

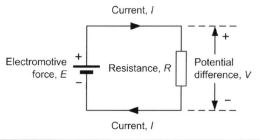

Figure 11.9 A simple circuit to illustrate the relationship between voltage, V, current, I, and resistance, R. Note that the direction of conventional current flow is from positive to negative.

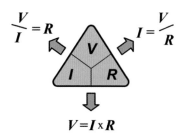

Figure 11.10 Triangle showing the relationship between V, I and R.

Example 11.11

A 12Ω resistor is connected to a 6V battery. What current will flow in the resistor?

Here we must use the relationship $I = \dfrac{V}{R}$ where $V = 6V$ and $R = 12Ω$.

Thus $I = \dfrac{6}{12} = 0.5A$

Example 11.12

A current of 0.1A flows in a 56Ω resistor. What voltage drop will appear across the resistor?

Here we must use $V = IR$ where $I = 0.1A$ and $R = 56Ω$.

From which $V = IR = 0.1 \times 56 = 5.6V$

Example 11.13

A voltage drop of 15V appears across a resistor in which a current of 20mA flows. What is the value of the resistance?

Here we must use $R = \dfrac{V}{I}$ where $V = 15V$ and $I = 0.02A$.

From which $R = \dfrac{15}{0.02} = 750Ω$

Test your knowledge 11.13

1 A resistor of 4Ω is connected across a 12V supply. What current will flow?

2 A voltage drop of 15V appears across a resistor when a current of 400mA flows in it. What is the value of the resistor?

3 A power supply is designed to supply an output of 15V, 0.75A. What value of load resistor would be required to test the power supply at its full rated output?

Energy and power

At first you may be a little confused about the difference between energy and power. Put simply, energy is the ability to do work while power is the rate at which work is done. In electrical circuits, energy is supplied by batteries or generators. It may also be stored in components such as capacitors and inductors. Electrical energy is converted into various other forms of energy by components such as resistors (producing heat), loudspeakers (producing sound) and light-emitting diodes (producing light).

The unit of energy is the joule, J. Power is the rate of use of energy and it is measured in watts, W. A power of 1W results from energy being used at the rate of 1J per second. Thus:

$$\text{Power}, P = \frac{W}{t}$$

where W is the energy in J, and t is the time in seconds, s.

The power in a circuit is equivalent to the product of voltage and current. Hence:

$$P = IV$$

where P is the power in watts (W), I is the current in amperes (A), and V is the voltage in volts (V). The formula may be arranged to make P, I or V the subject, as follows:

$$P = IV , \ I = \frac{P}{V} \ \text{and} \ V = \frac{P}{I}$$

The triangle shown in Figure 11.11 should help you remember these relationships.

The relationship, $P = IV$, may be combined with that which results from Ohm's Law ($V = IR$) to produce two further relationships. First, substituting for V gives:

$$P = I \times (IR) = P = I^2 R$$

Secondly, substituting for I gives:

$$P = \left(\frac{V}{R}\right) \times V = \frac{V^2}{R}$$

Example 11.14

A current of 1.5A flows in a circuit when it is supplied from a 3V battery. What power is dissipated?

Here we use $P = IV$ where $I = 1.5$A and $V = 3$V.

$$P = IV = 1.5 \times 3 = 4.5\text{W}$$

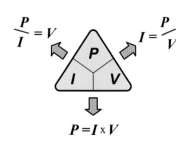

$$\frac{P}{I} = V \qquad I = \frac{P}{V}$$

$$P = I \times V$$

Figure 11.11 Triangle showing the relationship between P, I and V.

Example 11.15

A voltage drop of 4V appears across a resistor of 100Ω. What power is dissipated in the resistor?

Here we use $P = \dfrac{V^2}{R}$ where $V = 4V$ and $R = 100\Omega$.

$$P = \frac{V^2}{R} = \frac{4^2}{100} = \frac{16}{100} = 0.16 \text{ W}$$

> ### Key point
>
> Power is the rate at which energy is used. Power is measured in watts and 1W is equivalent to energy being used at the rate of one joule every second.

Test your knowledge 11.14

1. A current of 2A flows in a 22Ω resistor. What power is dissipated in the resistor?
2. A 9V battery supplies a circuit with a current of 75mA. What power is consumed by the circuit?
3. An electric motor is rated at 400W. How much energy is consumed by the motor if it runs continuously for 2 minutes?

Learning outcomes 11.11

Calculate resistors in series and parallel circuits

When more than one resistor (or *load*) is present in a circuit, they may be connected in series (Figure 11.12a), or in parallel (Figure 11.12b), or a combination of both methods.

Series resistors

The equivalent resistance, R_T, of two resistors connected in series (Figure 11.13) is given by:

$$R_T = R_1 + R_2$$

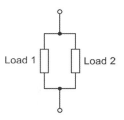

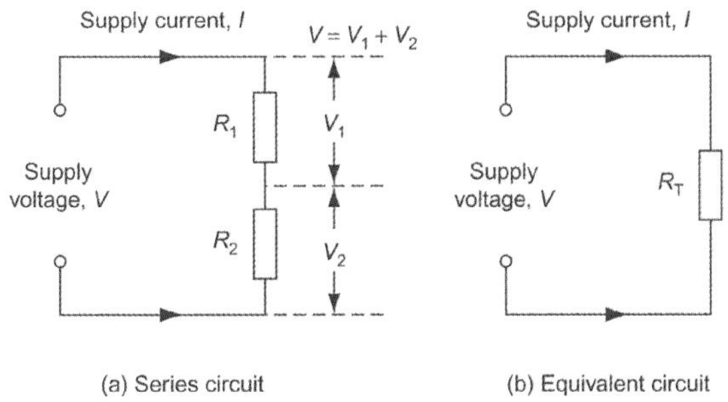

(a) Series connected

(b) Parallel connected

Figure 11.12 Series and parallel circuits.

(a) Series circuit   (b) Equivalent circuit

Figure 11.13 Resistor connected in series.

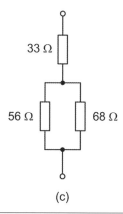

(a)

(b)

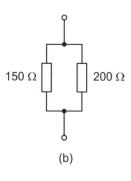

(c)

Figure 11.15 See Test your knowledge 11.15.

Parallel resistors

The equivalent resistance, R_T, of two resistors connected in parallel (Figure 11.14) is given by:

$$\frac{1}{R_T} = \frac{1}{R_1} + \frac{1}{R_2}$$

Note that this expression can be rearranged to give:

$$R_T = \frac{R_1 R_2}{R_1 + R_2}$$

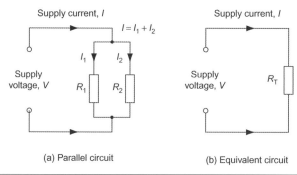

Figure 11.14 Resistors connected in parallel.

Example 11.16

Two 15Ω resistors are connected a) in series and b) in parallel. Determine the equivalent resistance of each arrangement.

a) In the series case, the equivalent resistance will be given by:

$$R_T = R_1 + R_2 = 15 + 15 = 30\Omega$$

b) In the parallel case, the equivalent resistance will be given by:

$$R_T = \frac{R_1 R_2}{R_1 + R_2} = \frac{15 \times 15}{15 + 15} = \frac{225}{30} = 7.5 \ \Omega$$

Test your knowledge 11.15

Determine the equivalent resistance of each of the circuits shown in Figure 11.15. *Hint*: Solve the parallel branch first for the circuit in (c).

Learning outcomes 11.12

Identify lines of flux within magnetic fields

Magnetism is the word we use to describe a force of attraction or repulsion that can exist between ferromagnetic materials such as iron, nickel, cobalt and their alloys. Every magnet has two poles, north and south. When the north and south poles of two magnets are placed in close proximity, they are attracted to one another. Conversely, when two north poles or two south poles are put together they will experience a force of repulsion. The force of attraction or repulsion will depend on the strength of the magnetism as well as how close the two magnets are. The closer they are the stronger the force will be.

When a magnet is broken into smaller pieces, each small piece, regardless of small it is, will have its own pair of north and south poles. This provides us with an important clue that helps us to realize that magnetism is caused by the behaviour of the elementary atomic particles within the material. Some ferromagnetic objects (such as bar magnets) are able to retain their magnetism indefinitely. They are referred to as *permanent magnets* and must be stored and handled correctly in order to avoid interference with any nearby magnetically sensitive objects.

Ferromagnetic materials can be turned into magnets by bringing them close to a permanent magnet. Figure 11.16a shows a ferromagnetic object that has not been influenced by the forces generated from another magnet. In this case, the individual tiny magnets are oriented in a random manner. Once the material is subject to the influence of another magnet, then these miniature magnets line up as shown in Figure 11.16b and the material itself becomes magnetic, with its own north and south poles.

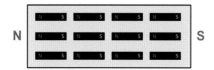

Miniature magnets randomly orientated. No overall magnetic effect

Miniature magnets aligned (material magnetized). Magnetic field lines produced.

(a)   (b)

Figure 11.16 The behaviour of ferromagnetic materials.

Magnetic fields

A magnetic field is the region in which the forces created by the magnet have influence. This field surrounds a magnet in all directions, being strongest at the end extremities of the magnet, known as the *poles.* Magnetic fields are mapped by an arrangement of lines that give an indication of strength and direction of the flux, as illustrated in Figure 11.17. When freely suspended in the horizontal plane a magnet aligns itself north–south parallel with the Earth's magnetic field. Because unlike poles attract, the north of the magnet aligns itself with the south magnetic pole of the Earth and the south pole of the magnet aligns itself with the Earth's north magnetic pole. This might be the opposite of what you would expect but it's important to be aware that the Earth's North Pole is actually a *magnetic* south while the Earth's South Pole is actually a *magnetic* north. Thus the red '*north-seeking*' end of a compass will point towards the Earth's *geographical* North Pole while the blue '*south-seeking*' end will point towards the Earth's *geographical* South Pole.

Permanent magnets should be carefully stored away from other magnetic components and any systems that might be affected by stray permanent fields. Furthermore, in order to ensure that a permanent magnet retains its magnetism it is usually advisable to store magnets in pairs using soft-iron keepers to link adjacent north and south poles. This arrangement ensures that there is a completely closed path for the magnetic flux produced by the magnets.

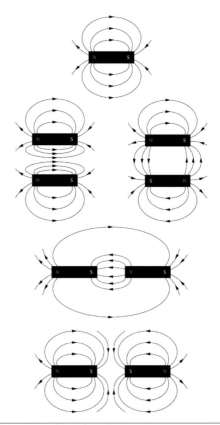

Figure 11.17 Field patterns for various bar magnet arrangements.

Figure 11.18 A magnetic compass with north- and south-seeking poles.

Test your knowledge 11.16

Give two examples of ferromagnetic materials.

Learning outcomes 11.13

Recognize the relationship between conductors, current, magnetic fields and relative movement

Whenever an electric current flows in a conductor a magnetic field is set up around the conductor in the form of concentric circles. The field is present along the whole length of the conductor and is strongest nearest to the conductor. As in the case of a permanent magnet, this field also has direction. The direction of the magnetic field is dependent on the direction of the current passing through the conductor and may be established using the *right-hand rule*, as shown in Figure 11.19. The arrows shown on the field lines mark the direction of a free north pole within the field. Notice how the lines of flux are concentric and form continuous circles around the wire.

If the right-hand thumb is pointing in the direction of current flow in the conductor the fingers will indicate the direction of the magnetic field. In a cross-sectional view of the conductor a point or dot (•) indicates that the current is flowing towards you (i.e. *out of* the page) and a cross (×) shows that the current is flowing away from you (i.e. *into* the page). This convention is easy to remember if you think of an arrow in flight, where the dot is the tip of the arrow and the cross is the feathers at the tail of the arrow.

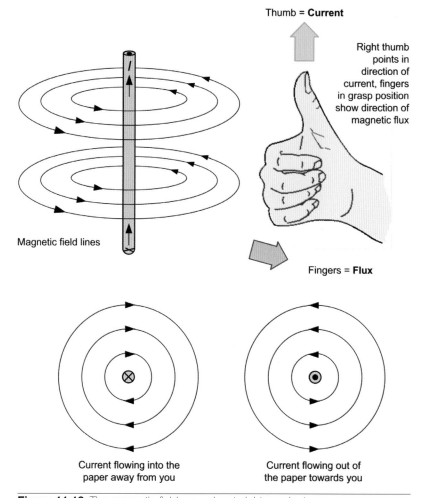

Figure 11.19 The magnetic field around a straight conductor.

Increasing the field strength

In order to increase the strength of the field, a conductor may be shaped into a loop (Figure 11.20) or coiled to form a solenoid (Figure 11.21) having a soft-iron (ferromagnetic) core. Note, in the latter case, how the field pattern is *exactly* the same as that which surrounds a bar magnet.

Electromagnets

As long as the magnetic core material is not saturated with magnetic flux the force of attraction or repulsion will be proportional to the current flowing in the coil. The magnetic flux (and lines of force) can then be turned on or off by simply interrupting the current flow. A typical application is illustrated in Figure 11.22.

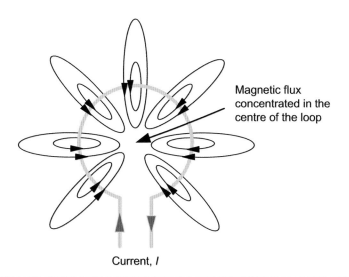

Magnetic flux concentrated in the centre of the loop

Current, *I*

Figure 11.20 Forming a conductor into a loop increases the strength of the magnetic field in the centre of the loop.

Figure 11.22 An electromagnet being used to separate ferromagnetic materials in a scrapyard.

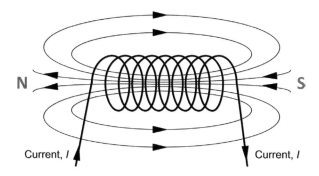

N   S

Current, *I*   Current, *I*

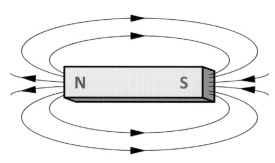

N   S

Figure 11.21 The magnetic field surrounding a solenoid coil resembles that of a permanent magnet.

Relays

Relays are electromechanical devices that consist of a coil of wire wound on a ferromagnetic core fitted with a moving armature. This, in turn, is linked to a set of switch contacts that make and break when the device is actuated. When sufficient current is applied to the coil of the relay the resulting magnetic field will cause the core to magnetize and the armature will be pulled towards the core and this in turn will open or close the relay's electrical contacts. A typical relay like the one shown in Figure 11.23 operates from 12V with less than 0.1A but is capable of switching voltages of up to 240V at up to 20A. One added bonus is that there is a very high degree of electrical isolation between the driving circuit (i.e. the circuit that supplies current to the coil) and the circuit that is being controlled.

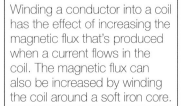

Key point

Whenever an electric current flows in a conductor a magnetic field is set up in the space surrounding the conductor. The field spreads out around the conductor in concentric circles, with the greatest density of magnetic flux nearest to the conductor.

Key point

Winding a conductor into a coil has the effect of increasing the magnetic flux that's produced when a current flows in the coil. The magnetic flux can also be increased by winding the coil around a soft iron core.

Figure 11.23 Internal arrangement of a typical relay showing the coil, armature and switch contacts.

Test your knowledge 11.17

Describe an application of electromagnetism.

Review questions

1. State the SI units for a) mass, b) length and c) time.

2. What is the weight of a glass panel having a mass of 2.2kg? (Take $g = 9.81\text{m/s}^2$)

3. A light alloy engine block weighs 375N. What is its mass? (Take $g = 9.81\text{m/s}^2$)

4. A 10mm square brass bar has a length of 450mm. If brass has a relative density of 8.5, what will the bar weigh?

5. A jack exerts a pressure of 1.25kPa. What force is applied if this pressure is exerted over an area of 0.04m^2?

6. Determine the turning moment produced by a lever having a length of 1.2m if a force of 15N is applied at right angles to the end of the lever.

7. Explain the difference between mass and weight.

8. Explain the difference between heat and temperature.

9. The aluminium wing spar of a light aircraft has a total length of 12m at 20°C. By how much will the wing spar increase in length at a temperature of 40°C?

10. Calculate the heat energy required to raise the temperature of a 7.5kg copper bar from 18°C to 53°C.

11. State **three** different ways in which heat can be transferred.

12. Calculate the horizontal force required to move a 450kg container over a floor if the coefficient of friction between the container and the floor is 0.475.

13. Identify the **three** different states of matter. Explain how these differ.

14. In an electric circuit how much charge is transferred if a current of 7.5A flows for 90 seconds.

15. A voltage drop of 27V appears across a 15Ω resistor. How much current will be flowing and what power will be dissipated in the resistor?

16. How much energy is consumed when a heating element rated at 500W is connected to a supply for 15 minutes?

17. A relay coil has a resistance of 700Ω. If the coil is connected to a 24V supply, determine the power supplied to the relay.

18. Explain, with the aid of a diagram, how an electromagnet works. Give one practical application of an electromagnet.

19. Resistors of 56Ω and 68Ω are connected a) in series and b) in parallel. Determine the equivalent resistance in each case.

20. If the two circuits in Question 19 are each connected to a 50V supply, determine:
    a) the total current supplied
    b) the current flowing in each resistor
    c) the voltage drop across each resistor
    d) the power dissipated by each resistor

Chapter checklist

Learning outcome	Page number
11.1 Recognize common SI units.	290
11.2 State the types of forces used in engineering.	292
11.3 Calculate moments and levers.	296
11.4 Calculate heat input and change in length.	299
11.5 Identify the modes of heat transfer.	302
11.6 List the causes of friction.	305
11.7 Identify how and why materials are selected with low frictional values.	306
11.8 Identify structures and states of matter.	307
11.9 Recognize the main principles of the basic theory of electricity.	309
11.10 Perform simple calculations using the basics of electricity.	311
11.11 Calculate resistors in series and parallel circuits.	315
11.12 Identify lines of flux within magnetic fields.	317
11.13 Recognize the relationship between conductors, current, magnetic fields and relative movement.	319

Appendix 1
Sample assessment

This assessment consists of 40 multiple-choice questions. Select one answer to each question. The numbers in square brackets relate to the syllabus group. Note that, within a real assessment, questions will appear in random order within each section.

Time allowed: 60 minutes

Section 1

1. Under which statutory regulations is it necessary for employers to provide face masks and goggles for use in a workshop where oxyacetylene welding is being carried out?
    a) The Provision and Use of Work Equipment Regulations
    b) The Personal Protective Equipment at Work Regulations
    c) The Workplace (Health, Safety and Welfare) Regulations
    d) The Management of Health and Safety at Work Regulations
    [1.1]

2. Under the Health and Safety at Work Act it is necessary for:
    a) employees to be given the same working hours every day
    b) employers to ensure that emergency exits are kept locked
    c) employers to provide on-site catering facilities for all employees
    d) employees to make proper use of equipment provided for their safety
    [1.1]

3. An interlocked machine guard ensures that:
    a) only trained operators can make use of the machine
    b) the machine can't be operated when the guard is open
    c) the work can only be loaded when the machine is running
    d) the work can only be unloaded when the machine is running
    [1.1]

4. In order to provide an effective working environment employers must:
   a) employ an on-site medical team
   b) ensure that it is safe and fit for purpose
   c) have good links with the local community
   d) provide facilities for exercise and recreation
   [1.2]

5. In an engineering company the layout of a new CNC workshop would be determined by:
   a) the Design Department
   b) the Human Resources Department
   c) the Production Department
   d) the Marketing Department
   [1.3]

6. Which one of the following is **not** a product associated with engineering companies that are active in the aerospace sector?
   a) Automatic blind landing systems
   b) Airport terminal buildings and hangars
   c) Ejector seats for use in military aircraft
   d) Gas turbine engines for use in helicopters
   [1.3]

7. Which one of the following could be classed as a threat to an engineering company?
   a) Interest rates fall to an all-time low
   b) A regional training centre opens nearby
   c) A major competitor is offering high salaries to new recruits
   d) The government introduces low-cost business development loans
   [1.3]

8. Which one of the following is **not** a direct consequence of industrial change on the requirements of the workforce?
   a) The importance of transferable skills
   b) The need for proficiency in the use of IT and ICT
   c) The need for ongoing training and development
   d) The ability to work with only minimal supervision
   [1.3]

9. Which one of the following correctly describes the role of an engineering technician?
   a) Applies proven techniques and procedures to solve engineering problems

b) Uses experience, leadership and management skills to direct the work of others

c) Carries out repetitive tasks associated with manual production and assembly

d) Responsible for recruitment and retention of engineering personnel at all levels

[1.3]

Section 2

10. Why are written notes made during an engineering team meeting?

a) To provide a detailed record of everything that was said

b) To assist senior management when allocating tasks and job roles

c) To summarize the discussion so that it can be reported in the local press

d) To provide a written record of agreed actions and those responsible for them

[2.1]

11. Which line type should take the highest priority when preparing an engineering drawing?

a) centre lines

b) dimensions

c) leader lines

d) visible outlines

[2.1]

12. Which one of the following is **not** a function of a computer operating system?

a) Providing a low-level interface to the computer's hardware

b) Providing an interface that will handle user input and output

c) Providing data on a wide variety of engineering components

d) Providing a file system that manages the storage of programs and data

[2.2]

13. Software that allows you to perform everyday tasks such as writing a letter, editing an image or sending an e-mail message is referred to as:

a) utility software

b) applications software

c)   programming software
d)   operating system software
[2.2]

14.   A database is a collection of records that you can access using:
a)   a query
b)   an e-mail
c)   a search engine
d)   a word processor
[2.2]

15.   The tool marked X in Figure A1.1 is:
a)   vice jaws
b)   vice shoes
c)   a vee-block
d)   an angle plate
[2.3]

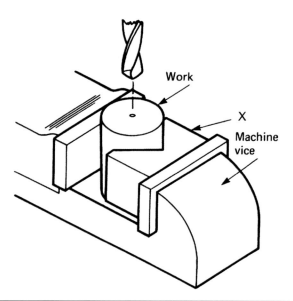

Figure A1.1 See Question 15.

16.   Figure A1.2 shows the tip of a drill bit. What do angles X and Y represent?
a)   X = rake angle; Y = clearance angle
b)   X = clearance angle; Y = rake angle
c)   X = taper angle; Y = clearance angle
d)   X = clearance angle; Y = taper angle
[2.3]

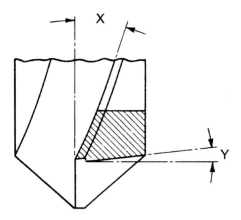

Figure A1.2 See Question 16.

17. Which one of the rivets in Figure A1.3 is a pan-head rivet?
   a) A
   b) B
   c) C
   d) D
   [2.3]

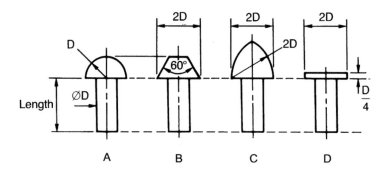

Figure A1.3 See Question 17.

18. You are using a digital multimeter to check the continuity of some patch leads. After a short time the display fades out and becomes impossible to read. What should you do next?
   a) Check, and if necessary, replace the display.
   b) Check, and if necessary, replace the battery.
   c) Return the instrument to the manufacturer and ask for a replacement.
   d) Return the instrument to the stores and use a battery and test lamp instead.
   [2.3]

19. When marking out, a surface plate provides:
    a)   a means of supporting a bench vice
    b)   a reference plane for measurements
    c)   a handy working area for cutting and filing
    d)   a way of storing the tools and instruments needed
    [2.4]

20. Before marking a workpiece out, the surface is often coated with a blue dye. The reason for this is:
    a)   It prevents rust, surface deposits and corrosion.
    b)   It improves the appearance of the finished part or component.
    c)   It avoids the need to remove oil, grease and finger marks from the surface.
    d)   It ensures that marked lines and dots contrast well with the background metal.
    [2.4]

21. Which one of the following is the reason for using a common datum when marking out?
    a)   It helps to avoid conversion errors.
    b)   It helps to avoid cumulative errors.
    c)   It reduces the number of drawings required.
    d)   It avoids the need for calipers, micrometers and other instruments.
    [2.4]

Section 3

22. Which one of the following materials is most suitable for use in the manufacture of a drill bit?
    a)   zinc-carbon alloy
    b)   low carbon steel
    c)   phosphor bronze
    d)   high carbon steel
    [3.1]

23. A quantity of finned heat dissipators are to be cut and machined from light alloy. Which one of the following forms of supply is most appropriate?
    a)   bar
    b)   plate

c) extruded

d) tube

[3.1]

24. Which one of the following pairs of properties would be essential for a material that's used to manufacture the mating electrical contacts of a connector fitted to an aircraft undercarriage?

a) low contact resistance and low corrosion resistance

b) low contact resistance and high corrosion resistance

c) high contact resistance and low corrosion resistance

d) high contact resistance and high corrosion resistance

[3.1]

25. When specifying the material for a threaded fastening, the load-bearing capability would be indicated by its:

a) density

b) hardness

c) brittleness

d) tensile strength

[3.2]

26. After heat treating, carbon steel will become:

a) stronger and harder

b) weaker and harder

c) stronger and softer

d) weaker and softer

[3.2]

27. Quenching involves:

a) hardening a workpiece by cold rolling

b) plunging a hot workpiece into a cold liquid

c) heating a workpiece until it becomes molten

d) grinding the surface of a workpiece until it is smooth

[3.2]

28. Express $\dfrac{3}{4} \times \dfrac{1}{2}$ as a decimal:

a) 0.125

b) 0.342

c) 0.375

d) 0.752

[3.3]

29. Which one of the following gives the area of the triangle shown in Figure A1.4?
    a) 43mm²
    b) 96mm²
    c) 225mm²
    d) 450mm²
    [3.3]

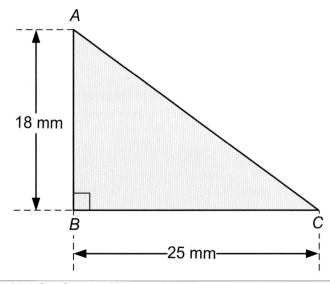

Figure A1.4 See Question 29.

30. If $x = \sqrt{16}$ and $y = 16^2$ which one of the following gives the values of x and y?
    a) $x = 8, y = 32$
    b) $x = 4, y = 32$
    c) $x = 8, y = 64$
    d) $x = 4, y = 256$
    [3.3]

31. If $A = \pi r^2$ which one of the following is TRUE?
    a) $r = \dfrac{A}{2\pi}$
    b) $r = A - \pi$
    c) $r = \dfrac{A^2}{\pi}$
    d) $r = \sqrt{\dfrac{A}{\pi}}$
    [3.3]

32. In Figure A1.5 which one of the following gives the length of side *AC*?
    a) 10mm
    b) 24.7mm
    c) 30mm
    d) 34.7mm
    [3.3]

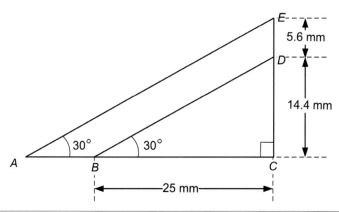

Figure A1.5 See Question 32.

33. A rectangular steel plate measures 152mm × 198mm. Which one of the following pairs of approximate values could be used to arrive at a result nearest to the true surface area of the plate?
    a) 150mm × 190mm
    b) 150mm × 200mm
    c) 160mm × 190mm
    d) 160mm × 200mm
    [3.3]

34. Which one of the following is the SI unit of pressure?
    a) coulomb, C
    b) kilogram, kg
    c) newton, N
    d) pascal, Pa
    [3.4]

35. Which one of the following gives the moment of the force shown in Figure A1.6?
    a) 28Nm
    b) 35Nm
    c) 63Nm
    d) 98Nm
    [3.4]

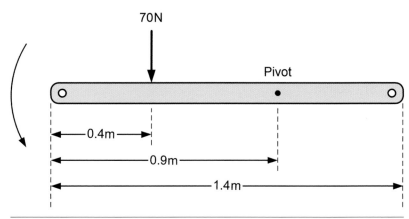

Figure A1.6 See Question 35.

36. Which one of the following is **not** a mode of heat transfer?
    a) conduction
    b) convection
    c) immersion
    d) radiation
    [3.4]

37. Which one of the following statements is **false**?
    a) Friction acts at right angles to a force that produces motion.
    b) The forces of weight and reaction act in opposite directions.
    c) Friction is caused by two rough surfaces sliding over one another.
    d) The coefficient of static friction is always greater than the coefficient of dynamic friction.
    [3.4]

38. In a metal conductor such as copper, electric charge is conveyed by:
    a) the movement of atoms
    b) the movement of neutrons
    c) the movement of electrons
    d) the movement of molecules
    [3.4]

39. A 12V battery is connected to a 48Ω resistor. Which one of the following gives the current that will be flowing in the resistor?
    a) 0.25A
    b) 0.5A

c) 4A

d) 6A

[3.4]

40. Which one of the magnetic field patterns shown in Figure A1.7 is correct?

a) A

b) B

c) C

d) D

[3.4]

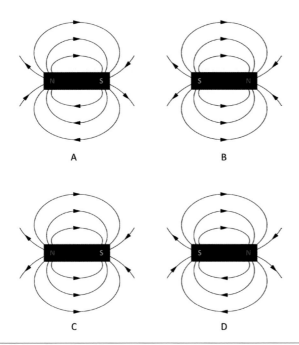

Figure A1.7 See Question 40.

Answers

1. b	2. d	3. b	4. b	5. c
6. b	7. c	8. d	9. a	10. d
11. d	12. c	13. b	14. a	15. c
16. a	17. b	18. b	19. b	20. d
21. b	22. d	23. c	24. b	25. d
26. a	27. b	28. c	29. c	30. d
31. d	32. c	33. b	34. d	35. b
36. c	37. a	38. c	39. a	40. c

Appendix 2
Using the Casio fx-83 calculator

The Casio fx-83 calculator is currently widely available and is highly recommended for use by engineering students at Level 2 and Level 3. The calculator has over 250 functions and it incorporates a 'natural display' that supports input and output using mathematical notation, such as fractions, roots etc.

Initializing the calculator

The keystrokes shown in Figure A2.2 can be used to return the calculator's settings and modes to their initial (default) values. Note that this operation will also clear any data from the calculator's memory. The fx-83's Clear menu is shown in Figure A2.3.

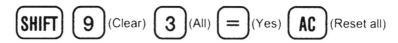

Figure A2.1 The Casio fx-83GT scientific calculator.

Figure A2.2 Key entry required to initialize the calculator.

```
Clear?
1:Setup        2:Memory
3:ALL
```

Figure A2.3 The fx-83's Clear menu will allow you to reset and initialize the calculator.

Setting the calculator mode

The fx-83GT provides three different calculation modes: COMP is used for general calculations, STAT is used for statistical calculations, and TABLE is used to generate a table of values based on a given expression. For Level 2, you will only need to use the COMP mode. The Mode menu is shown in Figure A2.4 and the keystrokes required to set the calculator to the COMP mode are shown in Figure A2.5.

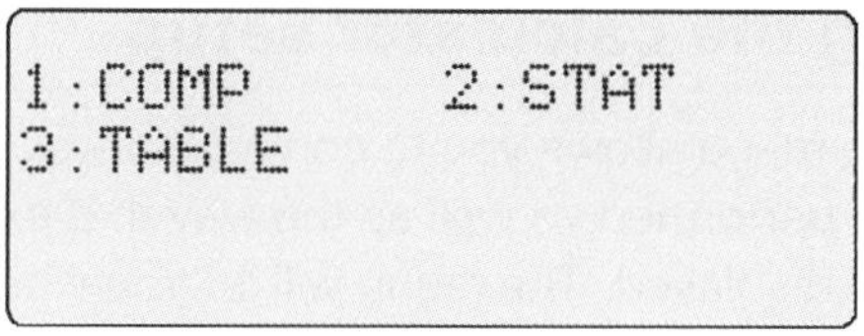

```
1:COMP      2:STAT
3:TABLE
```

Figure A2.4 The fx-83's Mode menu will allow you to change the calculator's mode.

 (COMP)

Figure A2.5 Key entry for setting the calculator to COMP mode.

Note: When you first power up the fx-83 calculator it will be in 'MathIO' mode. This may be awkward for use with many basic engineering calculations. To change the mode to the more conventional 'LineIO' mode, you need to press the SHIFT button followed by MODE and 2.

Table A2.1   The display symbols used on the fx-83.

Display	Meaning
S	The keypad has been shifted by pressing the SHIFT key. The keypad will unshift and this indicator will disappear when you press a key
A	The alpha input mode has been entered by pressing the ALPHA key. The alpha input mode will be cancelled and this indicator will disappear when you press a key
M	This indicates that a value has been stored in the calculator's memory
STO	The calculator is waiting for the input of a variable name. The calculator will then assign a value to this variable. The indication appears after you press SHIFT RCL STO
RCL	The calculator is waiting for the input of a variable name. The calculator will then recall the value of this variable. The indication appears after you press RCL
STAT	The calculator is in the statistical mode
D	The default unit for angles is degrees
R	The default unit for angles is radians
G	The default unit for angles is grads
FIX	The calculator has been set to display a fixed number of decimal places
SCI	The calculator has been set to display a number of significant digits
Math	The calculator has been set to the Maths input/output mode
▼▲	Calculation history is available and can be replayed (or there is more data above or below the display)
Disp	The display currently shows an intermediate result of a multi-statement calculation

Configuring the calculator setup

The fx-83 Setup menu allows you to control the way in which calculations are performed as well as the way that expressions are entered and displayed. The menu will let you work with a fixed number of decimal places (FIX), or with a fixed number of significant digits (SCI). You can also choose to use mathematical notation (MthIO) or conventional (line-based) notation (LineIO). The Setup menu is shown in Figure A2.7 whilst the keystrokes required to set the calculator to LineIO mode are shown in Figure A2.6.

Figure A2.6 Key entry for setting the calculator to COMP mode.

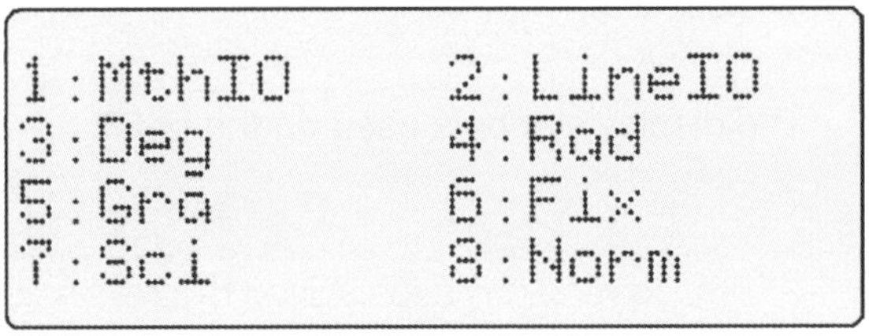

```
1:MthIO      2:LineIO
3:Deg        4:Rad
5:Gra        6:Fix
7:Sci        8:Norm
```

Figure A2.7 The fx-83's Mode menu will allow you to change the calculator's mode.

Appendix 3
Conversion table:
Inches to mm

Fractional inches			Decimal inches	Millimetres
		1/64	0.0156	0.396
	1/32		0.0313	0.793
		3/64	0.0469	1.190
1/16			0.0625	1.587
		5/64	0.0781	1.984
	3/32		0.0938	2.381
		7/64	0.1094	2.778
1/8			0.1250	3.175
		9/64	0.1406	3.571
	5/32		0.1563	3.968
		11/64	0.1719	4.365
3/16			0.1875	4.762
		13/64	0.2031	5.159
	7/32		0.2188	5.556
		15/64	0.2344	5.953
1/4			0.2500	6.350
		17/64	0.2656	6.746
	9/32		0.2813	7.143
		19/64	0.2969	7.540
5/16			0.3125	7.937
		21/64	0.3281	8.334
	11/32		0.3438	8.731
		23/64	0.3594	9.128
3/8			0.3750	9.525
		25.64	0.3906	9.921

Fractional inches			Decimal inches	Millimetres
	13/32		0.4063	10.318
		27/64	0.4219	10.715
7/16			0.4375	11.112
		29/64	0.4531	11.509
	15/32		0.4688	11.906
		31/64	0.4844	12.303
1/2			0.5000	12.700
		33/64	0.5156	13.096
	17/32		0.5313	13.493
		35/64	0.5469	13.890
9/16			0.5625	14.287
		37/64	0.5781	14.684
	19/32		0.5938	15.081
		39/64	0.6094	15.478
5/8			0.6250	15.875
		41/64	0.6406	16.271
	21/32		0.6563	16.668
		43/64	0.6719	17.065
11/16			0.6875	17.462
		45/64	0.7031	17.859
	22/32		0.7188	18.256
		47/64	0.7344	18.653
3/4			0.7500	19.050
		49/64	0.7656	19.446
	25/32		0.7813	19.843
		51/64	0.7969	20.240
13/16			0.8125	20.637
		53/64	0.8281	21.034
	27/32		0.8438	21.431
		55/64	0.8594	21.828
7/8			0.8750	22.225
		57/64	0.8906	22.621

Fractional inches			Decimal inches	Millimetres
	29/32		0.9063	23.018
		59/64	0.9219	23.415
15/16			0.9375	23.812
		61/64	0.9531	24.209
	31/32		0.9688	24.606
		63/64	0.9844	25.003
1			1.0000	25.400

Appendix 4
Answers to numerical 'Test your knowledge' questions

Chapter 10

1. 55

2. −45

a) −11

b) +12

c) −45

d) 77

e) −5

a), c), e), f) and g) are true

1. a) $\frac{3}{4}$, b) $2\frac{1}{4}$, c) $7\frac{3}{4}$

2. a) 3.25, b) 0.625, c) 4.1875

3. 7.75

4. $4\frac{3}{4}$

5. 4.25

6. $3\frac{3}{8}$

7. 12.5

8. 10m

9. 2.625 in

10. $\frac{3}{8}$ V

1. 75.2cm

2. 175.25kHz

3. 4.39s

4. 470µV

10.6   1. 8.202ft

2. 12.8016m

3. 129.162ft^2

4. 204.54l

5. 634.92lb

10.7   1. 5.167

2. 11, 10.5

3. 15, 17

10.8   1. 24

2. 3:1

3. 0.96kNm, 1.5:1

4. 30

5. 17

6. 0.7kg

10.9   1. $\rho = \dfrac{m}{V}$

2. 0.1292

10.10 a) 8m^2

b) 5m^2

c) 7m^2

d) 2.625m^2

e) 1525mm^2

10.11 1. 283.53mm^2

2. 34,557.52km

10.12 5.089m^3

10.13 1.672m^3

10.14 84,960mm^3

10.15 1. 34.335N

2. 14

10.16 0.875

10.17 1. a) 19.36, b) 729

2. a) 2.291, b) 44.452

3. 94.61, 269.88

10.18   1. $m = \dfrac{W}{g}$

2. $Z = \dfrac{V}{I}$

3. $C = \dfrac{1}{2\pi f X}$

4. $u = \dfrac{s}{t} - \dfrac{at}{2}$

5. $L = \dfrac{1}{f^2 4\pi^2 C}$

10.19   1. 11,111.1Pa

2. 3.6764m/s^2

10.20   75.398m/min; faster than expected

10.21   0.458mm/rev

10.22   1. 315 rev/min

2. 21.5m/min

10.23 $A = 45°$, $B = 60°$, $C = 75°$

10.24   25.2m^2

10.25   a) 2m^2

b) 2.5m^2

c) 2175mm^2

10.26   b) and c) are similar; angles are 45° and 105°; side is 81.25mm

10.27   1. 25

2. 25.185

3. 5 min

Chapter 11

11.1   1. a) kilogram, kg, b) kelvin, K, c) newton, N

2. I, ampere (A)

11.2   1. 40J/s or 40W

2. 10m/s^2

3. 150C

11.3   1. 3000kg/m^3; 4866N

2. 7.0318N

11.4   62.427kPa

11.5   75N

11.6   1. 302°F

2. -6.667°C

3. 89.6°F

11.7   23mm

11.8   11.148°C

11.9   8.338kN

11.12   1. 9kC

2. 0.18A

11.13   1. 3A

2. 37.5Ω

3. 20Ω

11.14   1. 88W

2. 0.675W

3. 48kJ

11.15   a) 60Ω

b) 85.7Ω

c) 63.7Ω

Index

Note: 'f' after a page number indicates a figure; 't' indicates a table.